Reviews of
103 Physiology, Biochemistry and Pharmacology

Editors
E. Habermann, Giessen · E. Helmreich, Würzburg
H. Holzer, Freiburg · R. Jung, Freiburg
R. J. Linden, Leeds · P. A. Miescher, Genève
J. Piiper, Göttingen · H. Rasmussen, New Haven
W. Singer, Frankfurt/M · U. Trendelenburg, Würzburg
K. Ullrich, Frankfurt/M · W. Vogt, Göttingen

With 45 Figures

Springer-Verlag
Berlin Heidelberg GmbH

ISBN 978-3-662-31042-7 ISBN 978-3-540-39420-4 (eBook)
DOI 10.1007/978-3-540-39420-4

Library of Congress-Catalog-Card Number 74-3674

Originally published by Springer-Verlag Berlin Heidelberg New York Tokyo in 1986
Softcover reprint of the hardcover 1st edition 1986

2127/3130-543210

Contents

Indexed in Current Contents

Rev. Physiol. Biochem. Pharmacol., Vol. 103

Recent Developments in Studies of the Supplementary Motor Area of Primates

MARIO WIESENDANGER

Contents

Institut de Physiologie, Université de Fribourg, Pérolles, CH Fribourg, Switzerland

1 Introduction

The notion of secondary motor areas, including the supplementary motor area (SMA[1]), arose from electrical-stimulation studies of the cerebral cortex (*Penfield* and *Welch* 1951; *Woolsey* et al. 1952). It is only in recent years that the SMA has emerged as a hierarchically more prominent cortical field than it had previously been thought to be. This change in view has been brought about by investigations of changes in regional cerebral blood flow (rCBF; *Lassen* and *Ingvar* 1972) and of 'readiness potentials', which occur up to 1 s before actual movement onset in man (*Kornhuber* and *Deecke* 1965). It was soon recognized that also areas outside the precentral motor cortex were activated when subjects were asked to perform repetitive movements. *Roland* et al. (1980a) and *Orgogozo* and *Larsen* (1979) attributed particular importance to the SMA, which, like the motor cortex, displayed an increased rCBF when patients performed a manual task. The SMA was found to be activated on both sides even during the mere 'mentation' of a complex sequence of finger movements. The discovery of these brain events relating to conscious motor performance led to the conclusion that the SMA functions as a 'supramotor' area which is engaged in the programming of sequences of voluntary movements; this inference was also supported by the fact that the 'readiness potentials' are

1 The terms MsII or MII have sometimes been used as synonyms for SMA

largest over the SMA (*Deecke* and *Kornhuber* 1978). All these studies on the human brain, together with some results of single-unit recordings in the SMA of trained monkeys, prompted *Eccles* (1982) to propose that ". . . in all voluntary movements the initial neural event is generated in the supplementary motor area of both sides".

This new view of a 'higher' role played by the SMA in initiating voluntary movements contrasts markedly with the more traditional view that the SMA controls the motor apparatus in parallel with the precentral motor cortex (*Woolsey* et al. 1952). The general objective of this review is therefore to examine the data on the SMA obtained from animal experiments and to determine if they support the new interpretation of the SMA as a *supramotor* area (rather than a *supplementary* motor area). The conceptual developments that emerged from experimental and clinical studies of the human brain will also be evaluated in the light of the proposition of *Eccles* (1982).

In the pursuit of the above general objective, the following specific questions will be addressed: (1) Is the SMA essentially 'upstream' from the precentral motor cortex, (2) does the SMA exert a direct and somatotopically organized control on the spinal cord, and (3) which inputs from sensory systems, from basal ganglia, and from the cerebellum influence the output of the SMA? These questions will be dealt with from the perspective of studies of structural relationships, the effects of lesions, and the characteristics of single-unit activity in monkeys trained to perform learned motor tasks. As will be seen in the detailed account, the conclusions are sometimes limited by a number of methodological and conceptual problems. These pertain firstly to some uncertainty on the exact spatial definition of the SMA. Secondly, many investigations, especially in human subjects, lack precision in the localization of lesions and of the sources of functional changes. Thirdly, where there is only a limited sample of single-unit recordings, the generality of the results is obviously limited.

In the present review, I adhere to the neutral reference of cytoarchitecture for defining the SMA, rather than using the original criterion derived from electrical stimulation. I postulate (as others, but not all, have done) that the SMA coincides with area 6 on the medial surface of the hemisphere (cf. Sect. 2.1 for details). This means that posteriorly, the SMA adjoins the hindlimb representation (and the tail representation in the monkey) of the motor cortex; ventrally, the anterior cingulate cortex of the limbic system; rostrally, the prefrontal association cortex; and laterally, the premotor cortex. Macroscopically, the SMA of the macaque brain occupies the superior frontal gyrus, including the roof of the cingulate sulcus. The posterior boundary is about at the level of the superior precentral dimple, the rostral boundary is about at the level where the upper branch of the arcuate sulcus ends (Fig. 2).

The present review focuses on recent studies and is not intended to cover systematically the older literature as well. A number of previous reviews have emphasized comparative aspects of the SMA and other motor fields (*Wiesendanger* et al. 1973; *Humphrey* 1979; *Wiesendanger* 1981a,b; *Brinkman* and *Porter* 1983).

2 Structural and Functional Relationships of the Supplementary Motor Area

2.1 Cytoarchitectonic Features

As outlined in the Introduction, it is presupposed that the SMA coincides with cytoarchitectonic area 6 on the medial wall of the hemispheres. It is therefore necessary to recall the essential features of area 6. Together with area 4, area 6 belongs to the 'agranular' type of cortex, which lacks an internal granular layer. The principal difference between the two fields is the presence in area 4 of giant Betz cells (which are particularly conspicuous in the medial cortex) and the lack of giant Betz cells in area 6. The cortex of area 6 is thinner than that of area 4, but the transition is gradual, and the criterion is therefore less reliable. The photographs originally published by *von Bonin* and *Bailey* (1947) and reproduced in Fig. 1 show nicely the respective cytoarchitectonic features of medial areas 4 and 6. Isolated clusters of giant Betz cells are sometimes present in front of the continuous sequence of Betz cells typical of area 4; it is therefore necessary to study a number of sections in order to establish a consistent boundary between the two areas. In our own studies (*Macpherson* et al. 1982a,b) we included these isolated Betz cell clusters into area 4. The boundary thus established corresponds fairly well to that of *von Bonin* and *Bailey* (1947) and is clearly more rostral than that of *Vogt* and *Vogt* (1919). Determining the transition into the granular cortex of the frontal lobe may also pose some problems because the granular layer may not appear sharp. *Walker* (1940) recognized a transitional 'dysgranular' strip of medial cortex, which he termed area 8B. The best-known cytoarchitectonic maps of the frontal-medial macaque brain are shown in Fig. 2. On the basis of our own experience (*Sessle* and *Wiesendanger* 1982; *Macpherson* et al. 1982a,b), I favour the map of *von Bonin* and *Bailey* (1947; Fig. 2B). From the functional point of view, area 6 is a large heterogenous cortical field. Anatomically, attempts have been made to subdivide area 6 into a rostral and a caudal portion, namely areas 6aα and 6aβ (*Vogt* and *Vogt* 1919) and areas FB and FC (*von Bonin* and *Bailey* 1947), respectively. Cytoarchitectonically, however, the differentiation is far from clear. But further hodological and functional findings will be discussed in

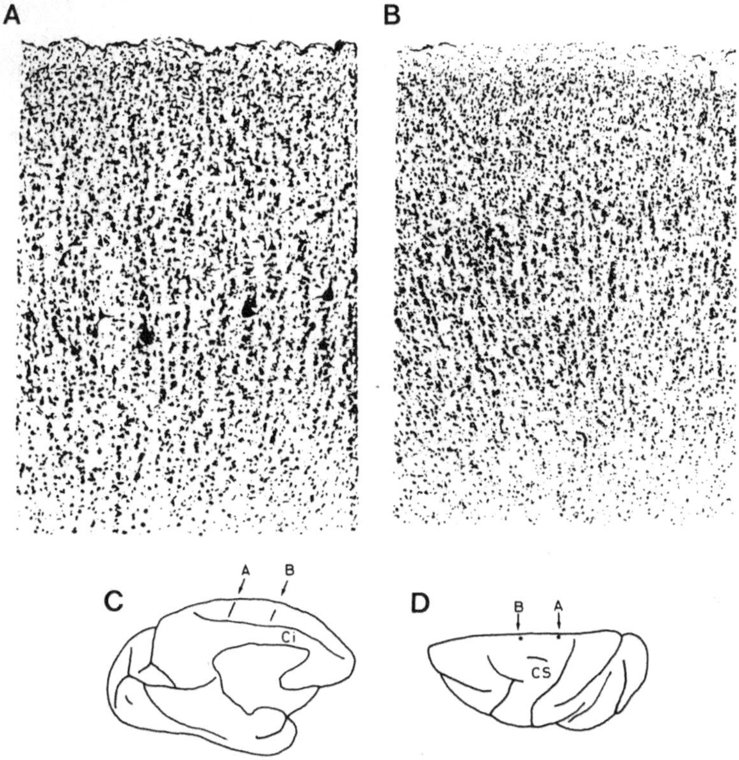

Fig. 1. A,B Histological sections of the macaque brain showing cytoarchitectonic features of the medial agranular cortex. **A** Medial area 4 of the motor cortex in the paracentral lobule. There are three giant Betz cells in this section. Note thickness of cortex and absence of an internal granular layer. **B** Medial area 6 (SMA). Note absence of giant Betz cells, lack of an internal granular layer, and lesser thickness of cortex than that of area 4. **C** Medial view and **D** dorsal view of the hemisphere, showing positions of the two sections (*A*, *B*). *Ci*, cingulate gyrus; *CS*, central sulcus. The microphotographs were reproduced from the original publication of *von Bonin* and *Bailey* (1947)

this review which indicate that distinction of a rostral and a caudal division of the SMA may indeed by justified.

Do the cytoarchitectonic features allow one to speculate on the evolutionary position of the SMA in relation to other cortical fields? *Sanides* (1968), in a cytoarchitectonic study of the squirrel monkey brain, used the term 'paralimbic premotor field' (Plpr) for medial area 6, in order to emphasize the similarities with the 'paralimbic belt' (the presence of a striking density of small pyramids in layer V). *Sanides* considers these similarities a sign of 'limbic influence' and speculates that the SMA may constitute, in an evolutionary sense, an 'old' motor field. As will be shown in Sect. 2.2.5, modern tracing studies have indeed revealed a close accociation of the SMA with the cingulate cortex. However, other considera-

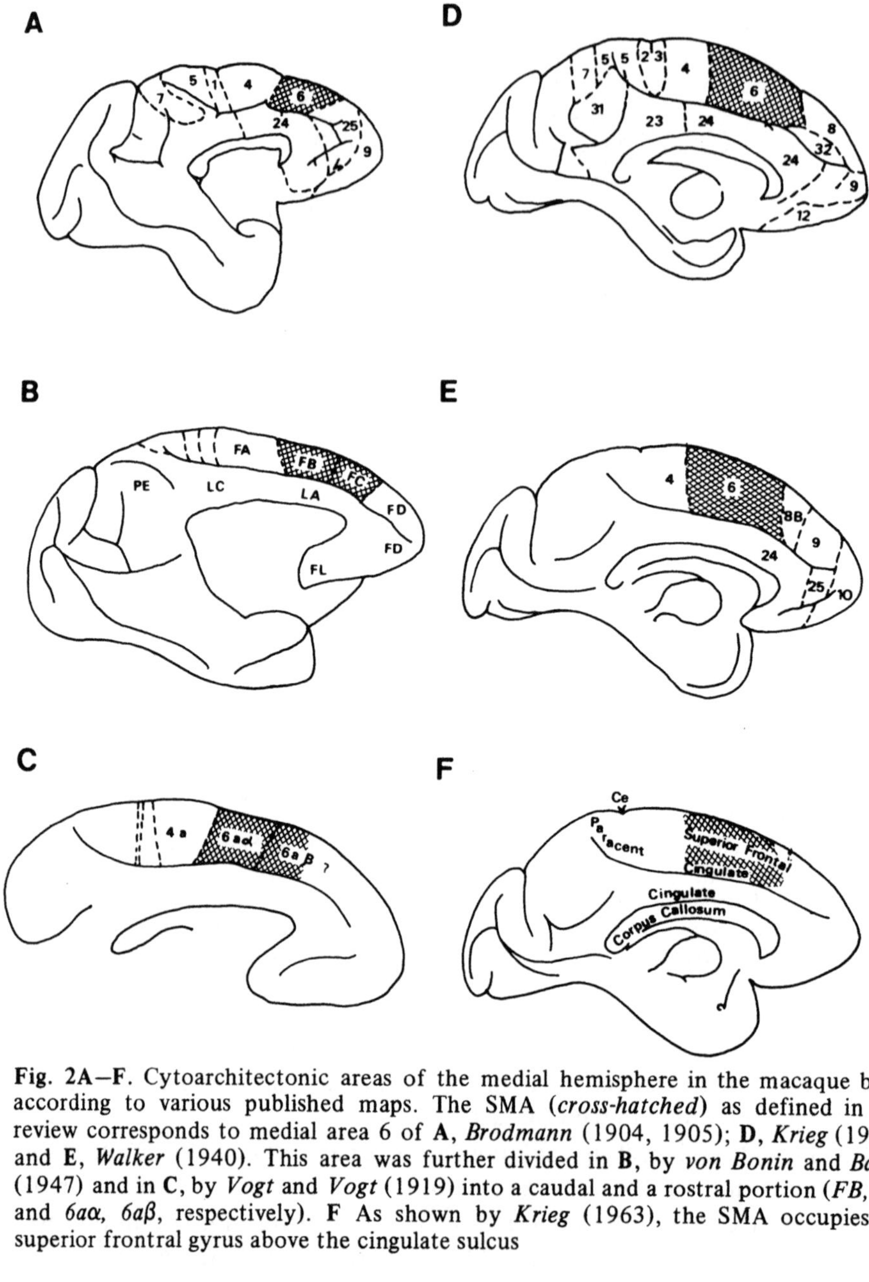

Fig. 2A–F. Cytoarchitectonic areas of the medial hemisphere in the macaque brain according to various published maps. The SMA (*cross-hatched*) as defined in this review corresponds to medial area 6 of **A**, *Brodmann* (1904, 1905); **D**, *Krieg* (1963); and **E**, *Walker* (1940). This area was further divided in **B**, by *von Bonin* and *Bailey* (1947) and in **C**, by *Vogt* and *Vogt* (1919) into a caudal and a rostral portion (*FB, FC,* and *6aα, 6aβ*, respectively). **F** As shown by *Krieg* (1963), the SMA occupies the superior frontral gyrus above the cingulate sulcus

tions do not lend support to the notion of an 'old' motor field. As will be detailed in several sections, many recent studies, especially those on the human brain, indicate that the SMA attained its full development in the human brain with attributes not present in the brain of subhuman primates.

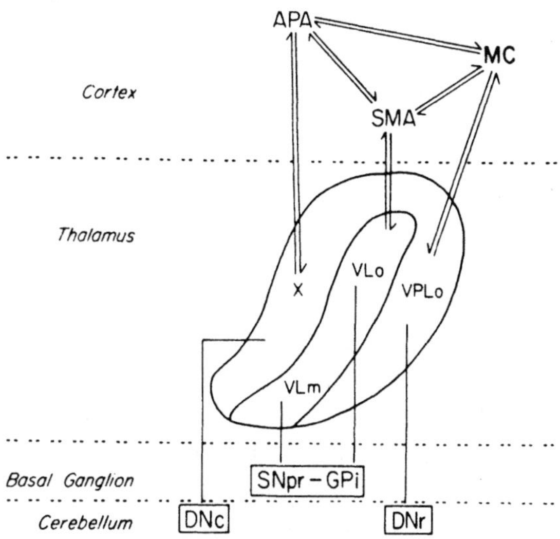

Fig. 3. Schematic representation of the principal thalamocortical relationships of the supplementary motor area (*SMA*), the arcuate premotor area (*APA*), and the motor cortex (*MC*) in the monkey. The main relation of the *SMA* is shown to be to the oral part of the ventrolateral nucleus (*VLo*), which in turn is the recipient of basal ganglia outflow. *APA* and *MC* are shown to receive their major inputs from the lateral cerebellum via thalamic areas *X*, of *Olszewski* (1952), and *VPLo*, the oral part of the ventroposterolateral nucleus, respectively. *VLm*, the medial part of the ventrolateral nucleus; *SNpr*, reticular part of substantia nigra; *GPi*, internal segment of globus pallidus; *DNc*, *DNr*, caudal and rostral portions of dentate nucleus. (*Schell* and *Strick* 1984)

2.2 Relationships to Subcortical Structures

2.2.1 Thalamocortical Connections and the Re-entrance Loops from the Basal Ganglia and from the Cerebellum to the SMA

It is well known that the large nuclear complex of the thalamus, which receives cerebellar, pallidal, and nigral afferents, is interconnected with motor fields of the cerebral cortex. The detailed work of *Jones* and coworkers and other relevant studies of the 'motor' thalamus have recently been reviewed in a series of papers (*Asanuma* et al. 1983a–d). Some scattered results from anterograde and retrograde tracing studies concerning the relationship of the thalamus to the SMA have been published by *Wiesendanger* et al. (1973), *Kievit* and *Kuypers* (1977), *Kalil* (1978), *Künzle* (1978), *Bowker* et al. (1979), and *Jürgens* (1984). However, *Schell* and *Strick* (1984) were the first to address specifically the problem concerning the relationship of the thalamus to the SMA. Their main result was that the SMA is dominated by thalamic inputs from subnucleus *VLo* (Fig. 3). The important significance of this observation is that *VLo* had been found to be the chief target structure of pallidal afferents (*Kim* et al.

1976). Therefore, *Schell* and *Strick* (1984) proposed that the SMA might be the key cortical structure receiving the signals from the basal ganglia. In contrast, lateral area 6, the premotor cortex, was found to receive its chief thalamic input from *Olszewski's* (1952) area X. The main relationships of the motor, premotor, and supplementary motor cortices to the thalamus are summarized in Fig. 3.

In our own independent anatomical study on the SMA hodology in the macaque (*Wiesendanger* and *Wiesendanger* 1985a) we have also mapped the reciprocal thalamocortical connections by means of autoradiography and histochemical labelling with horseradish peroxidase conjugated to the lectin wheat germ agglutinin (WGA-HRP). The reciprocal connection of the SMA with the VLo nucleus was confirmed. In addition, however, we also observed conspicuous anterograde and retrograde labelling in area X and in the VLc subnucleus after WGA-HRP injections into the rostral SMA (which was probably not investigated in the study of *Schell* and *Strick* 1984). It is fairly well established that these thalamic compartments in turn receive cerebellar afferents (see e.g. *Asanuma* et al. 1983b). Our results therefore support the inference that the SMA is a recipient, via the thalamus, not only of pallidal but also of cerebellar information. Furthermore, our results indicate that the anterior SMA and possibly also the posterior SMA are dominated by information from the cerebellum, whereas the middle portion appears to be more related to the basal ganglia outflow. Additional support for a cerebellothalamic input to the SMA was obtained from those cases in which the marker WGA-HRP was injected into the SMA. We discovered that, after a survival period of 5–6 days, WGA-HRP was transported transcellularly to label portions of the deep cerebellar nuclei, mainly the contralateral dentate nucleus (*Wiesendanger* and *Wiesendanger* 1985b). For comparison, WGA-HRP was also injected into the precentral cortex. The transcellular labelling pattern in the cerebellar nuclei was consistent with previous tracing studies of the cerebellothalamic link with the motor cortex (see *Asanuma* et al. 1983b). We therefore concluded that the transcellular labelling observed in our SMA cases was an additional and reliable sign of a linkage between the cerebellum and the SMA. Typical examples of thalamic and cerebellar-nuclear labelling following WGA-HRP injection into the SMA are illustrated in Fig. 4. Although it is our contention that the re-entrance loops both from the cerebellum and from the basal ganglia project back to the SMA, this does not imply that the two systems converge onto the same cell populations. It may well be that the two loops are kept separate at all levels, including the SMA, but this has not yet been tested at the cortical level.

In line with previous investigations on corticothalamic connections, we found the characteristics 'slabs' of thalamic labelling, which extend rostrocaudally over a distance of several millimetres. Sometimes the 'slabs'

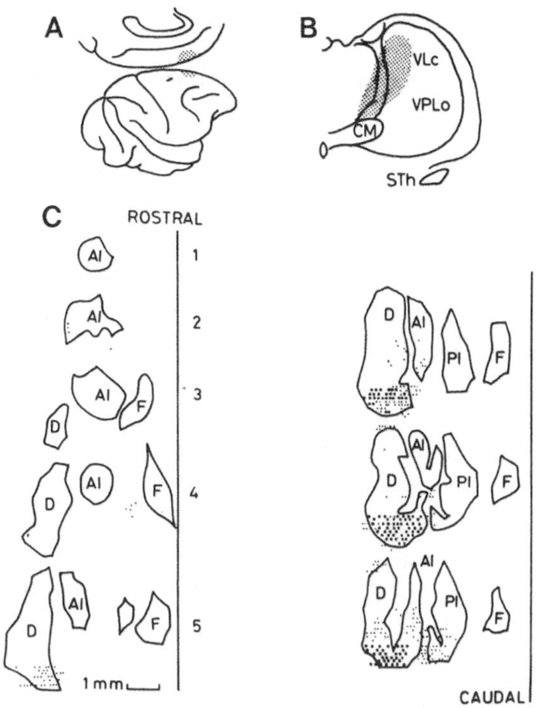

Fig. 4A–C. Retrograde thalamic and transcellular labelling in cerebellar nuclei following injection of the marker WGA-HRP into the rostral SMA in the monkey. **A** Lateral medial views of the cortex. **B** The labelled thalamic 'slab' includes the caudal part of the ventrolateral nucleus (*VLc*) and the nucleus centralis lateralis. *VPLo,* oral part of the ventroposterolateral nucleus; *CM,* centre median nucleus; *STh,* subthalamic nucleus. **C** Rostrocaudal sequence of transverse sections of the contralateral cerebellar nuclei. *Small dots* represent lightly labelled perikarya, *large dots* indicate relatively dense transcellular labelling of perikarya. *F,* fastigial nucleus; *AI* and *PI:* anterior and posterior portion of interpositus nucleus; *D,* dentate nucleus. (*Wiesendanger* and *Wiesendanger* 1985b)

are broken up in clusters extending rostrocaudally to form the thalamic 'rods', which have been described in detail by *Asanuma* et al. (1983c) and which are believed to constitute also functional modules of the thalamus. The other typical feature, which was first pointed out by *Kievit* and *Kuypers* (1977), was that the labelled territory transgressed subnuclear boundaries. In particular, it was noted that following WGA-HRP injections into the SMA, a number of nuclei were labelled which are outside the classical 'motor' thalamus (i.e. the ventralis anterior-ventralis lateralis [VA-VL] nuclear complex). Thus, prominent labelling was present in parts of intralaminar and adjacent nuclei, CL, Pc, lateral MD (Centralis lateralis, Paracentralis, Medialis dorsalis). That these nuclei may also be implicated in motor control has previously been suggested by *Jones* (1981), and physiological evidence has recently been adduced that single neurones of the intralaminar system subserve visuomotor functions (*Schlag-Rey* and

Schlag 1984). Another thalamic nucleus receiving afferents from cortical areas including the SMA is the centre-median-parafascicular complex; its possible function in motor control is unknown.

2.2.2 The Efferent Projection of the SMA to Subcortical Motor Structures

Together with other cortical areas, the SMA participates in the cortical projection to the corpus striatum. The caudate nucleus, and to a lesser extent, the putamen, receive a bilateral projection. Further projections are to the claustrum, the subthalamic nucleus, and the amygdalar nuclear complex (*Jones* et al. 1977; *Künzle* 1978a; *Jürgens* 1984; *Wiesendanger* and *Wiesendanger* 1984). Other (mainly ipsilateral) descending projections are to the parvocellular red nucleus, to the median portion of the pontine nuclei, and to the ipsilateral inferior olivary nucleus. A summary diagram of these connections is shown in Fig. 5. Some scattered and rather faint projections to the reticular formation have also been observed (*Palmer* et al. 1981; *Dhanarajan* et al. 1977; *Wiesendanger* and *Wiesendanger* 1984; *Jürgens* 1984; *Humphrey* et al. 1984). It may thus be concluded that SMA output is fed into the basal ganglia loop and the cerebellar loop. Access to the latter is provided by both climbing fibres and mossy fibres. It should be noted that the above target structures (with the possible exception of the reticular formation, which appears to receive only a limited input from the SMA) do not project down to the spinal cord. The influence of the SMA on the motor apparatus via subcortical relays is therefore indirect, involving complicated loops.

2.2.3 Extrathalamic Ascending Afferents to the SMA

The SMA shares with other cortical areas a number of ascending projection systems with partly known transmitters. After injections of a retrograde marker substance into the SMA, labelled neurones were seen in the claustrum and in the nucleus basalis of Meynert. Scattered labelled cells were found in the hypothalamus, in the prerubral area and in the central tegmental area. Numerous neurones were labelled bilaterally in the locus coeruleus. A moderate number of neurones of the raphe nuclei were also labelled, and scattered labelled cells were found in the nuclei annularis, centralis superior, dorsalis tegmenti, and ventralis tegmenti, and in the reticular formation (*Wiesendanger* and *Wiesendanger* 1984).

The following transmitter systems are thus likely to influence the SMA: cholinergic neurones of the basal forebrain nuclei, the noradrenergic neurones of the locus coeruleus, and the serotoninergic neurones of the raphe nuclei. Dopaminergic neurones are likely to contribute as well (*Goldman* and *Brown* 1981), although labelled cells were not seen in the

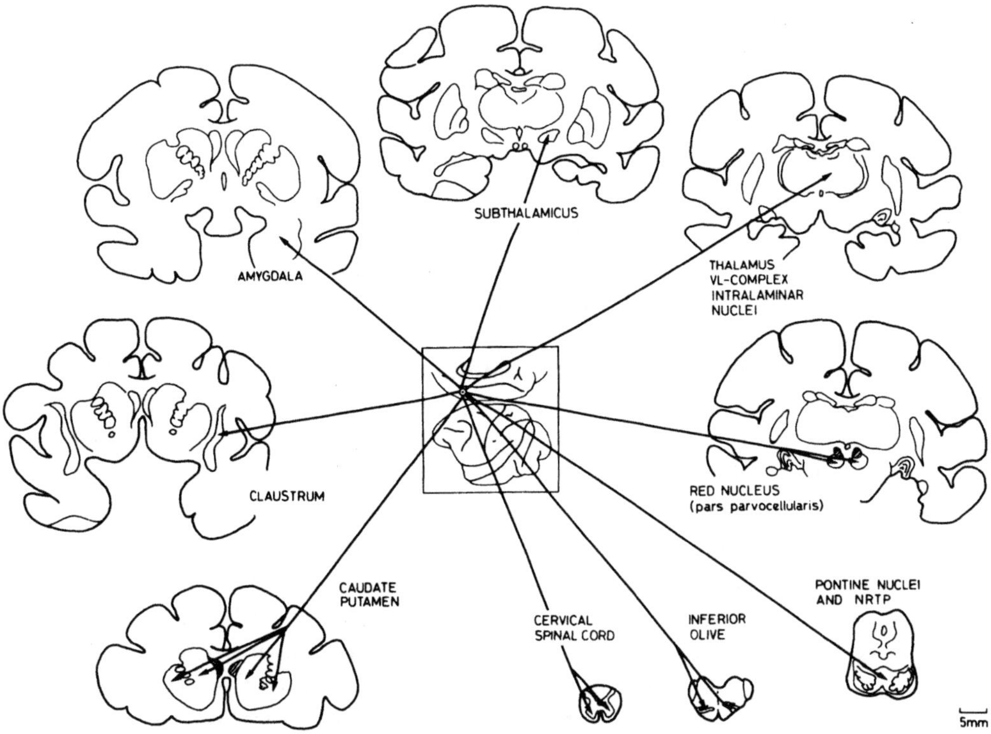

Fig. 5. Summary of major projections of the SMA to subcortical structures in *Macaca fascicularis.* Further minor projections (not illustrated) were observed in the following structures: septal nuclei, hypothalamic nuclei, interstitial nucleus of the posterior commissure, superior colliculus, and mesencephalic, pontine, and medullary reticular formation. *NRTP,* nucleus reticularis tegmenti pontis. (*Wiesendanger* and *Wiesendanger* 1984)

substantia nigra. Details of these projections to the SMA and their functional significance are unknown. One might speculate that, in analogy to the dopaminergic innervation of the corpus striatum, these systems exert an 'enabling function' on the cerebral cortex. Some support for this hypothesis comes from experiments of *Brozoski* et al. (1979), who showed that dysfunction of the ascending dopamine system (which also has a projection to the prefrontal cortex) provoked behavioural deficits similar to those observed after prefrontal lesions.

2.3 Corticocortical Relationships

2.3.1 Reciprocal Relations to the Precentral Motor Cortex

The powerful reciprocal connections of the SMA with the motor cortex have recently been reinvestigated in retrograde tracing studies by *Matsumura* and *Kubota* (1979) and by *Muakassa* and *Strick* (1979). The latter authors found that the caudal sector of the SMA projects to the precentral leg area; the middle sector, to the precentral hand area; and the rostral sector, to the precentral face area. This suggests that the SMA is indeed organized somatotopically, as previously advocated by *Woolsey* et al. (1952) on the basis of electrical-stimulation experiments. The question of somatotopy is, however, still a matter of debate, as will be further discussed in Sect. 2.4. Close analysis with anterograde tracing methods (autoradiography) revealed a corticocortical pattern of projection from the SMA consisting of rostrocaudal bands of silver grains which extended from the precentral area 4 to adjacent area 6 on the lateral surface (*Jürgens* 1984; *Wiesendanger* and *Wiesendanger* 1984). It is interesting to note that in both studies the projection from a single injection of labelled amino acids appeared to cover a considerable mediolateral area of the precentral cortex (with intervening bands that were clear of silver grains). The injections might have been too large to reveal a preferential somatotopic distribution such as had been found with retrograde tracing following injections of HRP into the motor cortex. Conversely, the motor cortex also projects back to the SMA. *Künzle* (1978b) found that the various subdivisions of the motor cortex showed an overlapping projection to the SMA, and he failed to see a precise point-to-point somatotopic organization.

Finally, the SMA also projects, via the corpus callosum, to the contralateral SMA and motor cortex (*Künzle* 1978a, and own observation).

2.3.2 Connections with Premotor Areas

The powerful links of the SMA with lateral area 6, the premotor cortex, alluded to in the foregoing section are also reciprocal (*Damasio* and *Van Hoesen* 1980; *Jürgens* 1984; *Wiesendanger* and *Wiesendanger* 1984). The frontal eye field of area 9 and the anterior cingulate cortex of area 24 are further cortical fields which have been associated with motor control functions and which are also reciprocally coupled with the SMA (*Muakassa* and *Strick* 1979; *Matsumura* and *Kubota* 1979; *Wiesendanger* and *Wiesendanger* 1984). Common to these premotor areas (in the broad sense) is that they contain a contingent of corticospinal neurones (see e.g. *Toyoshima* and *Sakai* 1982).

2.3.3 Relationships to the Sensory Cortex

Connections between motor and sensory areas have recently been demonstrated with anterograde and retrograde tracing methods by *Künzle* (1978), *Vogt* and *Pandya* (1978), *Jones* et al. (1978), and *Bowker* et al. (1979). The only sensory areas which have been found to project to the SMA are the somatosensory cortex (SI and SII) and the ‘supplementary sensory area' of medial area 5. In contrast to the SMA connections with motor fields, the SMA apparently does not project back to the somatosensory cortex.

In addition to the somatosensory input to the SMA, this area may relay more elaborate sensory information received from association areas (see below).

2.3.4 Relationships to Association Areas

Prominent reciprocal connections, which were seen in all cases investigated by *Jürgens* (1984), concerned the anterior and posterior cingulate cortex, i.e. areas 24 and 23, respectively. The projection from the cingulate cortex to the SMA was previously noted in anterograde tracing studies by *Pandya* et al. (1981), and the results are also in line with our own unpublished material.

Further reciprocal connections were seen with the frontal eye field and the prefrontal cortex of area 9 (*Jürgens* 1984), especially when the injections were placed in the anterior portions of the SMA (personal observations). Finally, there were relatively weak reciprocal connections with temporal and parietal association areas (*Jürgens* 1984; personal observations). In a most recent anterograde tracing study of the parietofrontal connections, *Petrides* and *Pandya* (1984) found that fibres destined to the SMA originate in the superior and middle parietal lobe including areas PE and PGm.

2.4 Anatomical and Electrophysiological Evidence for an SMA Connection with the Spinal Cord, and the Problem of Somatotopy

2.4.1 Corticospinal Neurones of the SMA

The uncertainties about the existence of corticospinal neurones in the SMA (cf. *Wiesendanger* 1981a) have now been clarified: with the method of retrograde labelling with the marker enzyme HRP, the presence of corticospinal neuronal cells has been unambiguously demonstrated by a number of investigators (*Biber* et al. 1978; *Murray* and *Coulter* 1981; *Toyoshima* and *Sakai* 1982; *Macpherson* et al. 1982a). The density of corticospinal neurones in the SMA is less than that in area 4, and their perikarya are, on the average, smaller (20.2 ± 4.1 μm diameter) than those of corticospinal neurones in the leg representation of area 4 (39.6 ± 9.1 μm; *Murray* and *Coulter* 1981). Camera lucida drawings of labelled corticospinal neurones of the SMA and of medial area 4 (containing the giant Betz cells) are shown for comparison in Fig. 6. The small size of corticospinal neurones in the SMA probably made it difficult to find a substantial

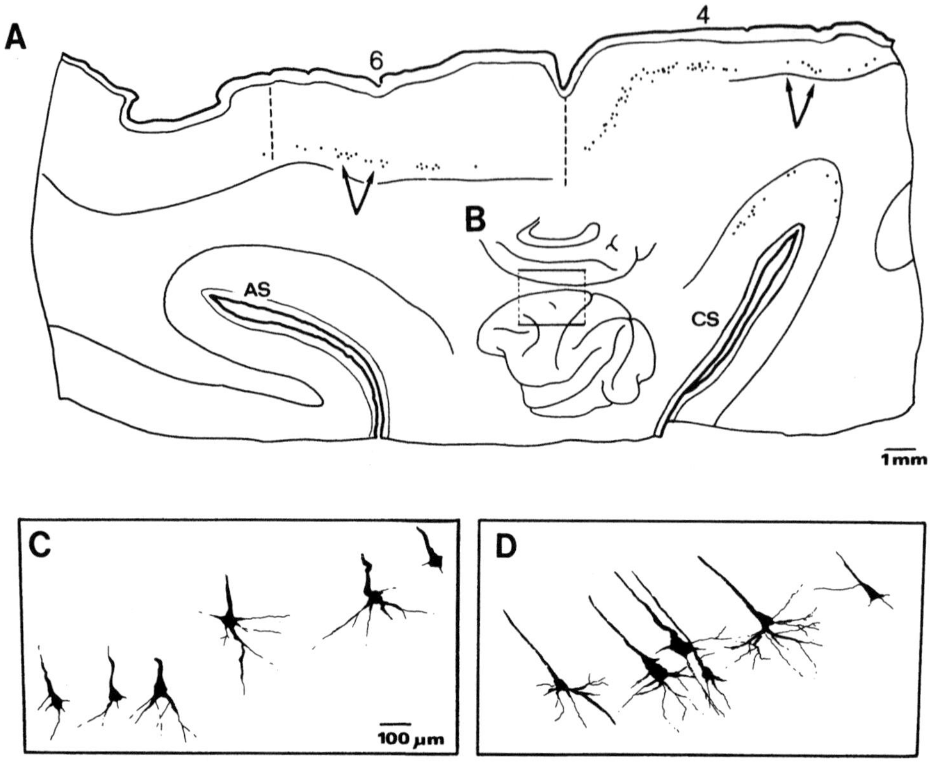

Fig. 6A–D. Labelling of corticospinal neurones in the medial agranular cortex of *Macaca fascicularis.* **A** Outline of a horizontal section. The corticospinal cells are marked by *dots. AS,* arcuate sulcus; *CS,* central sulcus. Note continuous row of corticospinal cells in medial area 6, opposite to the arcuate sulcus. *Arrows* indicate the locations of corticospinal cells shown in **C** and **D**. **B** (*inset*) Lateral and medial views of the monkey cortex, showing the orientation of the section. **C, D** Camera lucida drawings of retrogradely labelled corticospinal cells of the SMA (**C**), and of labelled corticospinal cells of medial area 4 (**D**), after large injection of HRP into the contralateral cervical cord. (Adapted from *Wiesendanger* and *Wiesendanger* 1984)

number of such cells by means of antidromic stimulation techniques (*Wiesendanger* et al. 1973; *Brinkman* and *Porter* 1979). However, an extensive search for descending corticofugal cells in the SMA with a set of stimulating electrodes at various levels of the neuraxis produced larger samples of corticospinal cells than were hitherto reported (*Macpherson* et al. 1982b). The distribution of conduction velocities of SMA and area 4 descending projection neurones is illustrated in Fig. 7.

2.4.2 Connections with Spinal Neurones, and the Effectiveness of Intracortical Stimulation in Activating Motoneurones

Whether corticospinal neurones of the monkey's SMA impinge directly on motoneurones is not firmly established. In a short note, *Cheema* et al.

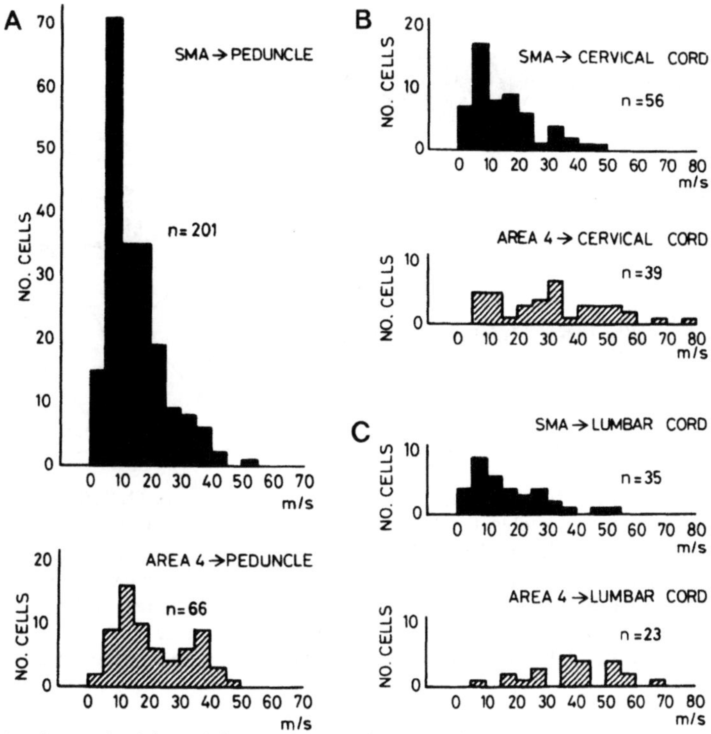

Fig. 7A–C. Conduction velocities of SMA and medial area 4 neurones in descending projection to the peduncle (**A**), the cervical cord (**B**), and the lumbar cord (**C**) in the monkey. Each histogram represents the caudal-most effective antidromic stimulus of the respective test. The majority of descending axons of SMA cells were backfired by the peduncular, but not by the spinal, stimulating electrode. Note low average conduction velocity of SMA cells. (*Macpherson* et al. 1982b)

(1983) reported that corticospinal fibres do not terminate in the motoneurone pool. On the other hand, *Brinkman* (1982) mentions in an abstract that some degenerating terminals were present in the motoneurone pool after lesions of the SMA.

The caudal zone of the SMA was clearly found to be microexcitable by intracortical stimulation in the awake monkey, which suggests a fairly close (but not necessarily monosynaptic) relationship to spinal motoneurones (*Macpherson* et al. 1982a). The mean latencies for responses in proximal and distal muscles were in a range similar to that of responses obtained with precentral microstimulation in the same animal (*Wiesendanger* et al. 1985a). Recently, we were able to demonstrate that single pulses of 30 μA or less produced a facilitation of motoneurones that were tonically active. The latencies of facilitation seen in the electromyograph (EMG) varied between 5.0 and 9.0 ms for forelimb muscles and between 10.0 and 14.5 ms for hindlimb muscles; these ranges were the same as for latencies obtained with single-pulse stimulation of the precentral cortex

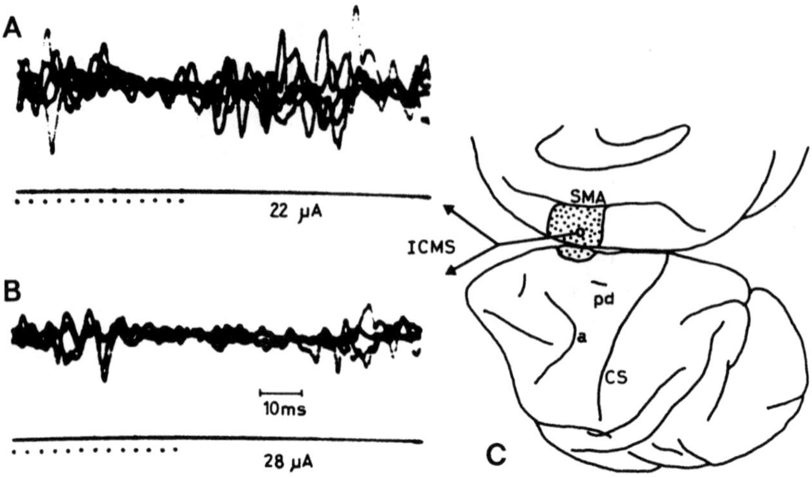

Fig. 8A–C. Effects of intracortical microstimulation (ICMS) of the SMA on contralateral trunk muscles (rectus muscle of abdomen). With stimulus intensities below 30 μA, localized, small twitch contractions were detected by palpation. **A, B** Electromyograph recordings revealed a tonic background activity, which was suppressed at a latency of about 20 ms. A rebound excitation occurred earlier with strong (**A**) than with light (**B**) background activity. **C** Approximate site of ICMS within the SMA (*dotted area*). (*Wiesendanger* and *Hummelsheim*, unpublished observations)

in the same animal (cf. also *Cheney* and *Fetz* 1985 and *Cheney* et al. 1985). Taken together, the above-mentioned anatomical and electrophysiological findings indicate that there is a fairly direct, possibly even monosynaptic, linkage of the caudal half of the SMA with spinal motoneurones.

2.4.3 The Preferential Distribution of Corticofugal Effects on Proximal Muscles

It was originally reported by *Penfield* and *Welch* (1951) that prolonged surface stimulation of the SMA activates preferentially proximal and trunk muscles. In fact, following ablation of the precentral cortex in monkeys, SMA effects were limited to proximal and trunk muscles. This finding was confirmed by *Wiesendanger* et al. (1973). With intracortical microstimulation, activation of shoulder muscles was by far the most common effect (*Macpherson* et al. 1982a). Movements of the hand could also be elicited, sometimes together with more proximal twitches, but this was rare.

Hindlimb effects, usually of the hip region, were also less prominent. It should be noted that, in our experimental situation, it was difficult to examine the trunk muscles. However, we repeatedly observed muscle contractions of abdominal muscles, such as illustrated in Fig. 8.

2.4.4 The Topical and Fine-Grained Organization of the SMA

Results of older studies obtained with prolonged surface stimulation revealed a coarse somatotopical organization of the SMA. The *simiusculus* representation given by *Woolsey* et al. (1952) is well known; it shows a rostrocaudal sequence of representations of the head, forelimb, and hindlimb, respectively. It is noteworthy that in an investigation in monkeys, *Penfield* and *Welch* (1951) observed a more complex picture, with hindlimb effects also seen in rostral parts of the SMA. Electrical surface stimulation (*Penfield* and *Welch* 1951) or intracortical stimulation (*Talairach* and *Bancaud* 1966) in neurological patients failed to reveal a rostrocaudal somatotopic pattern.

In our anatomical study (*Macpherson* et al. 1982a), labelled corticospinal neurones projecting to lumbosacral levels of the cord were more numerous in the caudal portion of the SMA, which was in agreement with *Murray* and *Coulter* (1981). However, a row of labelled cells was also found in the more rostral SMA after injections of the marker enzyme into the lumbar cord (especially in the upper bank of the cingulate sulcus). This observation fits well with the findings of *Penfield* and *Welch* (1951), who found hindlimb effects in deep portions including the rostral SMA, and with our own observations after intracortical microstimulation (*Macpherson* et al. 1982a). Taken together, our previous and some new results of microstimulation and an analysis of receptive fields in SMA neurones show a slight rostrocaudal shift with much overlap, between the representations of the forelimb and of the hindlimb, as illustrated in the histograms of Fig. 9.

The problem of somatotopy in the SMA was further investigated electrophysiologically (*Macpherson* et al. 1982b). To this end, single neurones of the SMA were identified by the method of antidromic stimulation, and the caudal-most projection was established by means of a series of stimulating electrodes along the rostrocaudal extent of the spinal cord. Corticospinal cells were classified as cervicothoracic if they responded to antidromic invasion by the high or low cervical electrode, but not be one stimulating more caudally. Corticospinal neurones also responding to stimuli applied to the lumbar cord were termed lumbosacral. Histological reconstruction of the locations of these corticospinal neurones of the SMA revealed that the cervicothoracic projection neurones tended to be located in the mesial wall of the SMA, whereas the lumbosacral projection neurones were above or below the former group. The results of microstimulation and of recording from corticospinal neurones in monkeys are thus more in line with the observations of *Penfield* and *Welch* (1951) than with those of *Woolsey* and co-workers (1952). However, the problem of somatotopy is still not settled. Using another approach, *Brinkman* and *Porter* (1979) and *Tanji* and *Kurata* (1982) have found that activity of

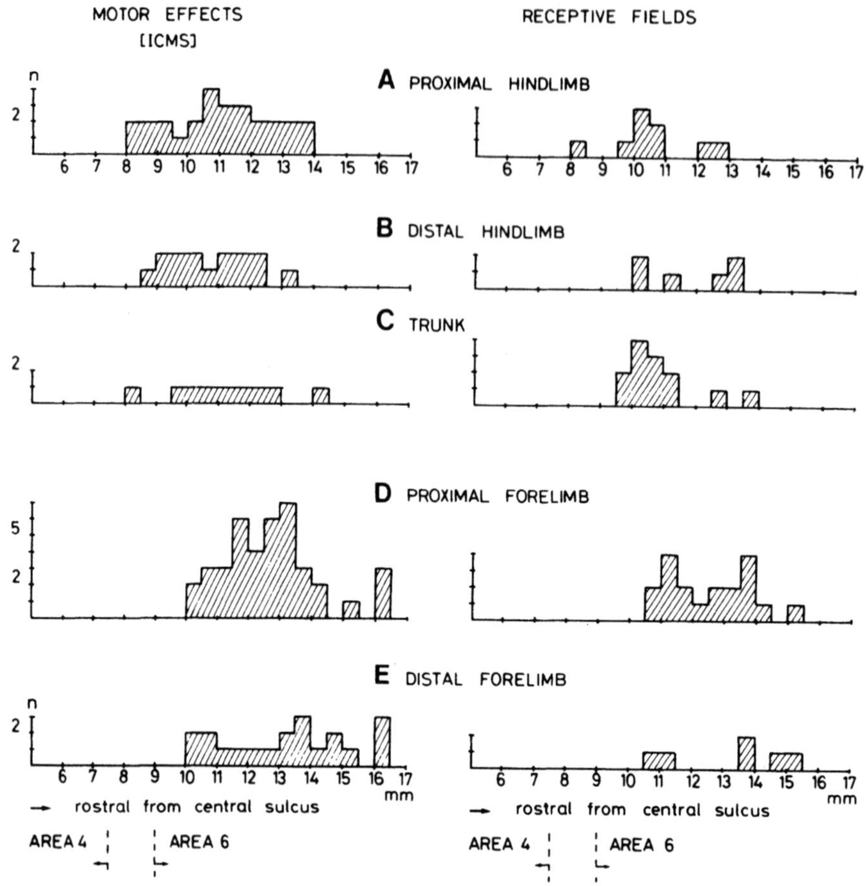

Fig. 9A–E. Rostrocaudal somatotopic representations in the SMA as revealed by motor responses to intracortical microstimulation (ICMS) and by analyses of receptive fields of SMA units in the awake monkey. The responses are plotted on *abscissae* with the scale indicating the distances between the microelectrode track sites and the central sulcus at its juncture with the midline. The transition between medial areas 4 and 6 was between 7.5 and 9.0 mm rostrally from the central sulcus. (After *Macpherson* et al. 1982a and *Hummelsheim, Bianchetti,* and *Wiesendanger,* new, unpublished observations)

single neurones recorded in the awake animal was related either to forelimb or to hindlimb movements and that the neurones related to the former were situated in more rostral tracks in the SMA than those related to the latter. Likewise the units related to proximal limb movements tended to be in more caudal tracks than cells related to distal movements (cf. also *Tanji* and *Kurata* 1979). There are, however, a number of problems in deducing somatotopical principles from such studies. First, with the complex task used by *Brinkman* and *Porter* (1979), it is difficult to be sure that 'distal' movements implicate exclusively distal muscles (no EMG recordings of proximal muscles were made in these studies). The problem of selection may also play a role. In the report by *Tanji* and *Kurata* (1982) their Fig. 2 shows only a few recording tracks that were situated caudally to the arcuate spur (their O coordinate) and were related to forelimb movements; these tracks were in their hindlimb zone. However, Fig. 2 of *Tanji* and *Kurata* (1979) shows many recordings at the same caudal levels that had revealed neuronal activity related to proximal forelimb movements.

On the basis of all current evidence, it may be concluded that some gradient of rostrocaudal organization in the connection of the SMA to forelimb and lindlimb muscles is consistent with the Woolsey scheme. However, all investigators seem to agree that there is much more overlap of representation than in the precentral cortex. Added difficulty in interpretation is caused by the fact that the hindlimb representation of area 4 is contiguous with mesial area 6 (and thus with Woolsey's hindlimb zone of the SMA).

A solution of the conflict may be found in more detailed investigations of the *fine-grained organization* of the SMA. In preliminary experiments, we have investigated more closely the microstimulation effects by recording the responses from multiple forelimb muscles with seven chronically implanted EMG electrodes (*Wiesendanger* et al. 1985b). It was found that widely distributed muscles could sometimes be activated from a given efferent zone. Thus, at stimulation thresholds of 30 μA or less, a combination of muscles were coactivated, such as hand extensors, and triceps brachii and brachioradialis muscles; or deltoid, brachioradialis, pectoralis and triceps branchii muscles. Such divergent efferent zones have not been observed in area 4. This could mean that the efferent zones of the SMA are more intermingled (and activated simultaneously by the spread of excitation), or that there is a higher degree of collateralization of descending axons of the SMA than for those of area 4. Both possibilities, the intermingling of efferent zones and a high degree of collateralization, may be of considerable functional significance, as will be further discussed in Sect. 6.3.

2.4.5 Indirect Influence of the SMA on the Spinal Cord

The indirect influences of the SMA involve complicated, multisynaptic loops, as detailed in Sect. 2.2.2. The projection to subcortical motor centres that in turn project further down to the spinal cord is weak. Thus, the SMA is linked with the parvocellular portion of the red nucleus, which does not project to the spinal cord. The SMA projection to the reticular formation is rather sparse. Nevertheless, it is possible that the SMA-reticulospinal system may have contributed to the motor effects observed after repeated stimulation of the cortical surface at high intensities. According to *Kuypers* (1981), area 6 (including its medial protion) controls proximal and trunk muscles mainly via subcortical motor centres in the tegmentum.

Finally, a corticocortical link via the precentral cortex also has to be considered. There is some anatomical evidence that the efferent projection of the SMA to the motor cortex is organized somatotopically in that the rostral portion of the SMA connects mainly with the precentral face area; the middle portion, with the precentral arm area; and the caudal portion, with the precentral hindlimb area (*Muakassa* and *Strick* 1979). Could it be that the somatotopy observed in some of the single-unit studies in performing monkeys is determined essentially by the corticocortical links?

2.5 Sensory Inputs to the SMA

As outlined in Sect. 2.3.3, the SMA receives prominent (non-reciprocal) projections from somatosensory cortical areas (SI, SII, medial area 5), which suggests that the SMA makes use of somesthetic signals to generate its output. The question also arises whether other modalities may influence SMA neurones, although no direct afferents from primary visual or auditory areas have been demonstrated.

In man, somatosensory evoked potentials (SEPs) have indeed been recorded from the region of the SMA (*Libet* et al. 1975). The latencies were long (125–165 ms), but the responses were considered to be generated locally, as evidenced by intracortical recordings. The same authors also recorded long-latency responses to visual and auditory stimuli. *Foit* et al. (1980) recorded similar late evoked responses over the region of the SMA when stimulating electrically the median nerve.

In lightly anaesthetized monkeys, electrical nerve stimulation elicited field potentials at latencies of 10–30 ms for forelimb nerves and of 15–30 ms for hindlimb nerves (*Wiesendanger* et al. 1973). These potentials were more prominent in the caudal portion of the SMA. It was furthermore noted that forelimb nerve stimulation and hindlimb nerve stimulation were equally effective in eliciting field potentials in Woolsey's forelimb area of the SMA. Many single units tested with these electrical stimuli were not responsive, but a sample of 23 neurones were discovered which reacted consistently to peripheral nerve stimulation, often displaying a convergence from two or more peripheral nerves (*Wiesendanger* et al. 1973). *Brinkman* and *Porter* (1979) reported that in awake monkeys the somatosensory responses to mechanical stimuli were weak, variable, and complex, with unclear receptive fields. *Smith* (1979), who tested 24 SMA cells for response to mechanical stimulation, found that 13 cells reacted to limb displacements, light pressure, or muscle tapping. In an abstract, *Gallouin* and *Albe-Fessard* (1973) reported that somatosensory responses were recorded in SMA sites from where motor effects were elicited by intracortical stimulation. These observations have been confirmed in recent

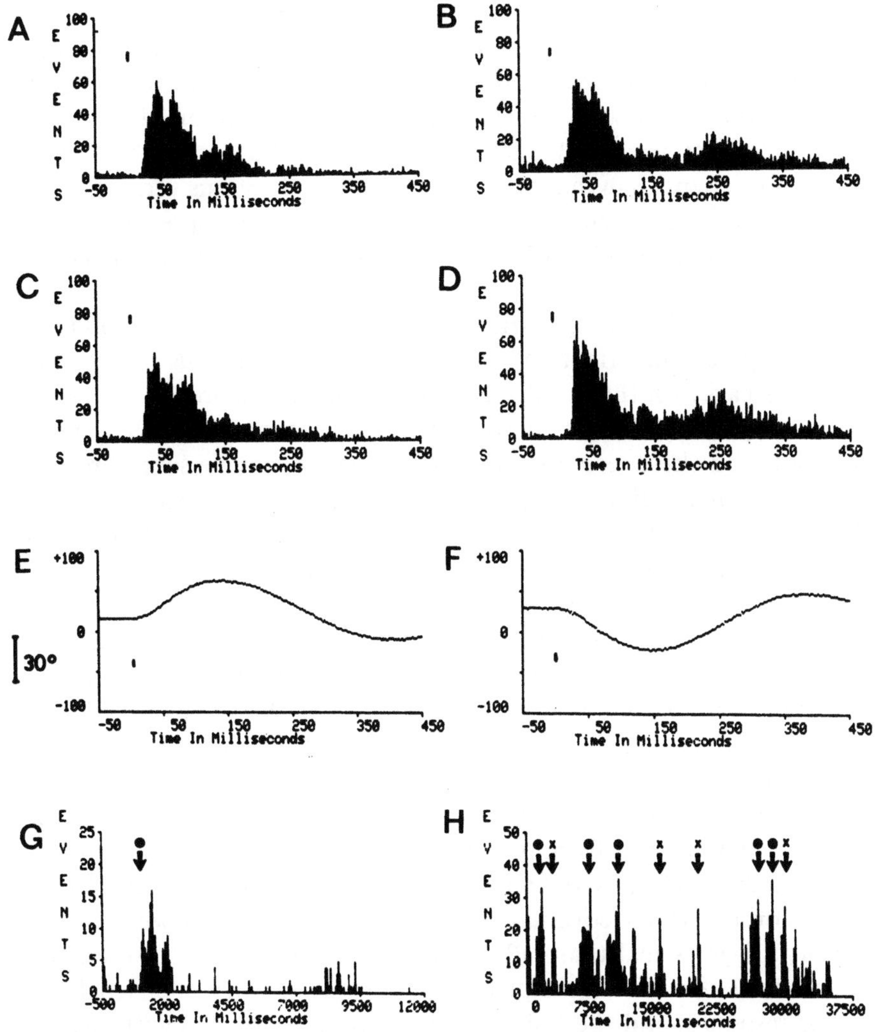

Fig. 10A–H. Somatosensory responses of a SMA neurone. **A–D** Peristimulus time histograms, each obtained during 50 arm displacements either flexing (**A, C**) or extending (**B, D**) the elbow in a passive but awake monkey. The neurone responded consistently with high-frequency discharges and with short latencies to these perturbations, applied at regular (**A, B**), or at irregular (**C, D**), intervals. **E, F** Averaged displacements with flexion (**E**) and extension (**F**). **G, H** The same neurone was also activated when the monkey reached for food with the contralateral arm (●) or when licking its fingers (*x*). Time zero indicated by ***small vertical bars***. Bin width 1 ms. (*Hummelsheim* and *Wiesendanger*, unpublished observations)

experiments on awake monkeys subjected to passive limb perturbations (*Wiesendanger* et al. 1985a). A sample of 102 neurones were collected which reacted consistently to elbow flexion, elbow extension, or both. Examples of response patterns are shown in Fig. 10. Responses of these neurones were recorded in the region of the SMA from which positive microstimulation effects were obtained (cf. Sect. 2.4.2). When precentral neurones of the arm focus were analysed in the same way, similar response patterns were seen with shortest latencies in the order of 10–15 ms in both areas. The peak of the latency histogram of SMA cells was, however, more to the right than that of precentral cells (20–30 ms versus 12–25 ms, respectively). The latencies of the somatosensory responses of neurones in the SMA are compatible with the assumption of a transmission via somatosensory or motor cortex.

These finding seemingly differ from those of *Wise* and *Tanji* (1981), who compared the effects of foot displacements on cells in the hindlimb zone of area 4 with those on cells of adjoining medial area 6. These authors found very few cells which were modulated during the displacement of the foot, and these responses were weak. There may be several reasons for this discrepancy: first, in our experiments, we were specifically seeking the perturbation-sensitive cells and found them to be clustered together with many unresponsive cells between them (on the average, 16% of the neurones were responsive); secondly, we concentrated our search in those regions of the SMA where positive microstimulation effects in shoulder and arm muscles had been observed; and thirdly, in our experience, occurrences of hindlimb motor effects in the SMA were relatively rare.

The question naturally arises whether the sensory responsiveness of SMA neurones may contribute to the sensory guidance or triggering of movements. Recordings of single units in a behavioural context clearly indicate that SMA cells may use sensory cue signals to generate output. Thus, in the study of *Tanji* and *Kurata* (1982), neural activity of the SMA was recorded which covaried with hand movements as well as with the sensory trigger signal that called for the movement. Somaesthetic, visual, or auditory stimuli were equally effective as sensory cue signals for the motor response, and the SMA activity was better related with the sensory signal than with the motor response. Furthermore, the authors found that the response time of the SMA units to the visual or the auditory cue signals was shorter than in precentral units tested under comparable conditions. Interestingly, the 'sensory response' of SMA cells was closely dependent on whether a movement followed the triggering signal.

The above results thus indicate strongly that sensory cue signals may indeed play an important role in generating SMA output and movements. On these gounds it is suggested that the SMA is not exclusively involved in self-initiated movements. It must be noted that, whereas the role of visual signals for modulating premotor neurones in visual tracking tasks is reasonably well established (see e.g., *Kubota* and *Hamada* 1979), the importance of such signals for the SMA has, to my knowledge, not been explored in a visual tracking task.

2.6 Synthesis of SMA Hodology

The SMA hodology as evidenced by anatomical, and to some extent also by electrophysiological, results may be subsumed under three categories: the connections with motor, sensory, and limbic systems. Regarding the first class of connections, there are many paths through which the SMA may gain *access to motor centres.* A contingent of output neurones project directly to the spinal cord which of course provides the most direct link with the motor apparatus. However, compared with the motor cortex, this contingent of corticospinal neurones is relatively weak. All other output is by far more indirect, either via corticocortical fibres impinging on the motor cortex, or via the two large re-entrance loops which implicate the cerebellum and the basal ganglia. It may thus be inferred that much of the activity generated in the SMA exerts its influence on the spinal cord after a considerable time lapse.

The second class of SMA relationships is relevant to the question whether the SMA is implicated in the *sensory guidance of movements.* The prevalent sensory information that appears to reach the SMA is from somatosensory areas, but further processed sensory information may reach the SMA via cortical association areas.

A third class of connections concerns the striking *relationship with the cingulate cortex.* In a very general way, this relationship may indicate that the SMA output is also conditional on 'internal' signals (motivation, drive).

3 Single-Unit Activity in Monkeys Performing Learned Movements

3.1 The Timing of Movement-Related Cell Discharges in the SMA

The first reports on single-unit data obtained in monkeys performing a motor task appeared in the late 1970s. *Brinkman* and *Porter* (1976, 1979) studied SMA neurones while trained monkeys performed a complex forelimb task. *Tanji* and *Taniguchi* (1978) were interested in correlations with sensory cue signals triggering a simple arm movement, and *Smith* (1979) investigated the implication of SMA neurones in the performance of a precision grip. Clearly, the objectives and the experimental approach in these studies were quite different, and this may to some extent explain the different outcomes. I shall first deal with the question whether neurones in the SMA that were found to correlate with some aspects of movement behave differently from cells in the motor cortex. The specific question posed in this section is whether SMA neurones change their firing rate before movement execution and perhaps even before the activation of cells in the motor cortex. Obviously, this question is difficult to answer if comparable data are not available, i.e. results obtained from the SMA and from the motor cortex with the same behavioural paradigm and, if possible, on the same animal by the same investigator. *Brinkman* and *Porter* (1979, 1983) concluded that their observations of a lead time of 150–200 ms with respect to movement onset (as determined from movies) and of only 120–180 ms with respect to EMG onset "... suggest a role for SMA in the initiation of movements". This should, however, not be interpreted in a strict sense, i.e. to mean that SMA activity occurs *before* that of the motor cortex. First, the authors have not compared in detail the timing of SMA cells and motor cortex cells (the latter had been recorded with the same paradigm by *Lemon* et al. 1976). Secondly, the authors do not show histograms of rate changes with respect to EMG or movement onset time; their EMG recordings were only from muscles of the forearm, but it is likely that EMG changes in trunk and proximal muscles occurred much earlier. Finally, the periresponse histograms were plotted at a bin width of 60 ms, which makes it difficult to detect subtle timing differences. The authors were of course aware of this and also mention that "... the experiments were not appropriate for examining the exact temporal relationships of neuronal discharge to movement performance...". This point is dealt with in detail because in the subsequent literature (e.g., *Roland* et al. 1980a; *Eccles* 1982) the reported lead time of 150–200 ms was taken as an important argument for the proposed role of the SMA in movement *initiation*.

Tanji and *Kurata* (1979) tested both a shoulder movement and a finger movement (key press) in the same animals. These movements were elicited by visual cue signals in a reaction time situation. The mean lead times of SMA cell activity were 134 ms and 234 ms before key press and shoulder movement, respectively. These lead times are indeed remarkably long, but unfortunately, no comparative data for the motor cortex are available from this test.

The data of *Smith* (1979), on the other hand, suggest that SMA neurones are recruited rather late in the performance of a precision grip. Movement onset times were distributed broadly, from –470 ms to +500 ms, with a peak at EMG onset time (time 0). In fact, the majority of the analysed 59 SMA cells changed their firing rate *after* movement onset. Comparison of these results with those on the motor cortex that were obtained previously in the same paradigm (*Smith* et al. 1975) led to the conclusion that SMA cells were recruited, on the average, 100–150 ms *later* than motor-cortical neurones! It should be noted that in such a seemingly distal motor task, proximal and trunk muscles may be coactivated before the distal muscle, which would render the mean interval between cell activity changes and the earliest EMG onset time even larger. It might be argued that this late recruitment of SMA cells was due to the fact that the trials were self-initiated and not triggered by a sensory stimulus as in the study of *Tanji* and *Kurata* (1979, 1982). However, the conditioned movements studied by *Brinkman* and *Porter* (1979) were also self-initiated. In all these considerations, it should be remembered, however, that the recruitment time for a given movement pattern is highly variable from neurone to neurone, as was convincingly shown for cells of the motor cortex (*Porter* and *Lewis* 1975). The difficulties arising when one attempts to interpret the timing of cell discharges in a reaction time situation has perhaps best been discussed by *Fetz* (1981): "A fundamental problem in attempting to demonstrate serial activation of different motor centers is that any particular region, including motoneuron pools, contains cells that are recruited over diverse times, making it difficult to interpret the relative onset times of particular cells in different regions. Moreover, since the duration of most movements greatly exceeds the conduction time between centers, recurrent loops could be 'traversed' many times during a single response; thus, the conceptually appealing notion that initiation of movement involves sequential activation seems difficult to prove." In this respect, it should be recalled that the motor cortex has a prominent feedback projection to the SMA (cf. Sect. 2.3.1). It is also very likely that the timing of SMA cell activation depends in each case on the specific inputs and outputs of a given neurone. For example, the firing of descending output cells may be closely coupled with a particular movement whereas that of other neurones may be more intimately coupled with the 'sensory' stimulus.

Rolls (1983) has addressed the question of temporal order of neural events during natural feeding behaviour. In several brain structures, the earliest changes of neuronal activity were related mainly to visual and gustatory guiding signals. In an abstract, *Thorpe* and *Rolls* (1982) have now reported that 73 of 125 cells recorded in the SMA were also activated in relation to feeding movements, such as reaching for food or licking. Sometimes these changes in cell firing occurred before the movements, but in all cases the latencies of the response to the eliciting stimuli were long (170–180 ms). It was concluded from these and previous data that SMA neurones were not the leading cells in the initiation of the particular type of motor behaviour; hypothalamic and other cells preceded those of the SMA, and it was thought that the SMA is more involved in execution than in preparation of movements.

In summary, the data thus far tend to indicate that a large proportion of SMA neurones change their discharge rate before an observable movement. However, it would be premature to conclude from these data that the SMA is more intimately involved in movement initiation than the precentral motor cortex.

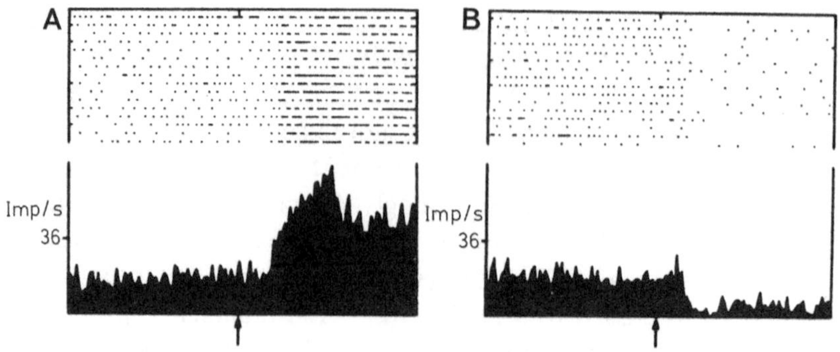

Fig. 11A,B. Responses of an instruction-related neurone of the SMA in a monkey trained to perform two different tasks, depending on previously presented instruction signals. Note increased discharge following a 'pull' instruction (**A**) and decreased discharge following a 'push' instruction (**B**). The discharge rate of this neurone was not time-locked with actual movement. *Arrows* indicate appearance of instruction signal. The display spans 1024 ms before, and 1024 ms after, the instruction signal. Bin width: 16 ms. (*Tanji* et al. 1980)

3.2 Covariance with Instruction Signals

Inspection of time histograms of neuronal discharges that were correlated with a conditioned movement often revealed that the onset and the peak of the discharges were not as prominent as commonly seen in neurones of the motor cortex (*Tanji* and *Kurata* 1982; *Brinkman* and *Porter* 1979, 1983). *Tanji* and co-workers were therefore particularly interested in the question whether SMA-neural activity may reflect the 'handling' of instruction signals for a forthcoming movement rather than the coding of a motor command (*Tanji* and *Taniguchi* 1978; *Tanji* et al. 1980; *Tanji* and *Kurata* 1982, 1983, 1985; *Kurata* and *Tanji* 1985). Accordingly, these authors chose the strategy of studying the simplest possible movement that may be conditioned by a series of different sensory cue signals. With a similar paradigm, previously used by *Tanji* and *Evarts* (1976), monkeys were trained to respond by either pushing or pulling a manipulandum, depending on whether a green or a red instruction signal appeared before the 'go' signal. From "many hundreds of neurones" recorded in the SMA, *Tanji* et al. (1980) collected a sample of 201 neurones related to the instruction signal. Slightly less than half of the neurones were specifically related to either the green or the red instruction signal. An example of the responses of such an instruction-related neurone is shown in Fig. 11. Significantly, these neurones showed no relation to the movement per se. The hypothesis which motivated these experiments is that the SMA might 'instruct' the motor cortex how to respond to a forthcoming triggering event, thus 'setting' the responsiveness of the motor-cortical cells.

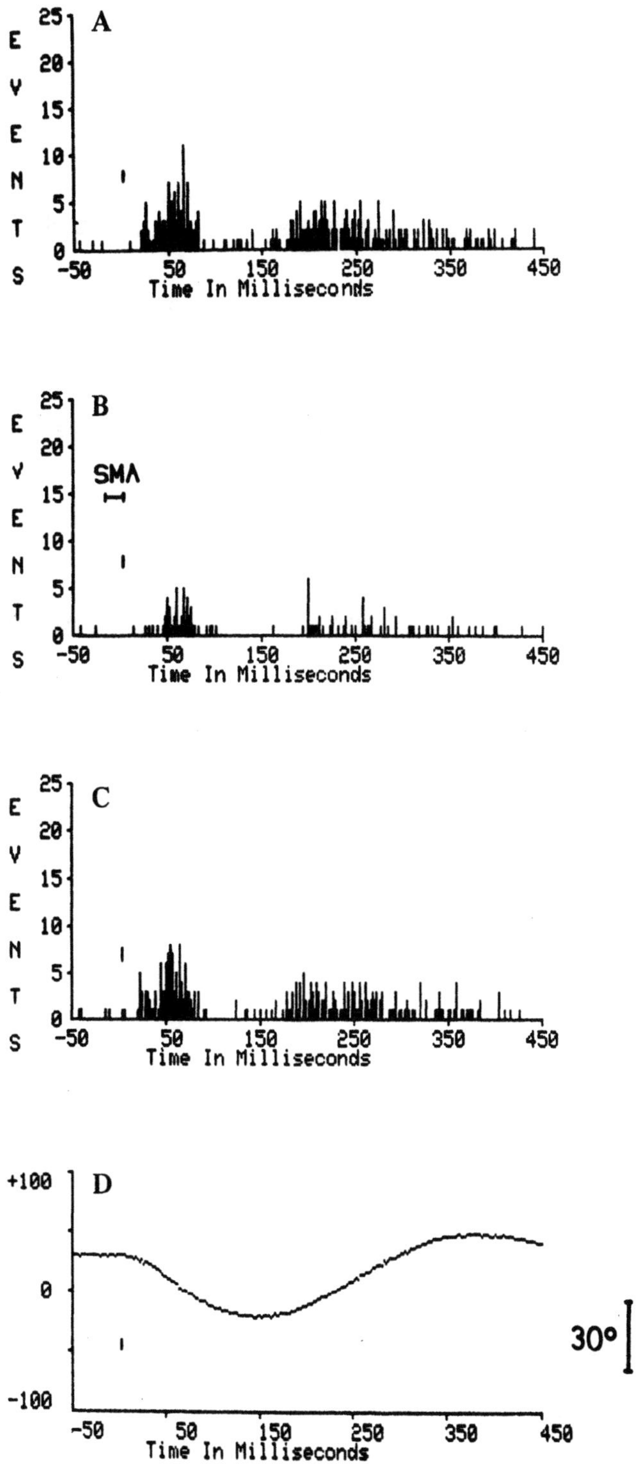

Fig. 12A–D. Modulation of an area 4 neurone by conditioning stimulation of the SMA in the awake monkey. The peristimulus time histograms show the responses of a neurone in the precentral arm area to 50 imposed elbow extensions. After control recordings (**A**), conditioning electrical stimulus (intracortical train, 330 Hz, 0.2 ms, about 200 μA) applied to the SMA reduced the sensory response to elbow extension (**B**). **C** Subsequent control without conditioning stimulation. Bin width: 1 ms. *Small vertical bars* indicate time zero. **D** Averaged displacement of elbow extensions. (*Wiesendanger* and *Hummelsheim*, unpublished observations)

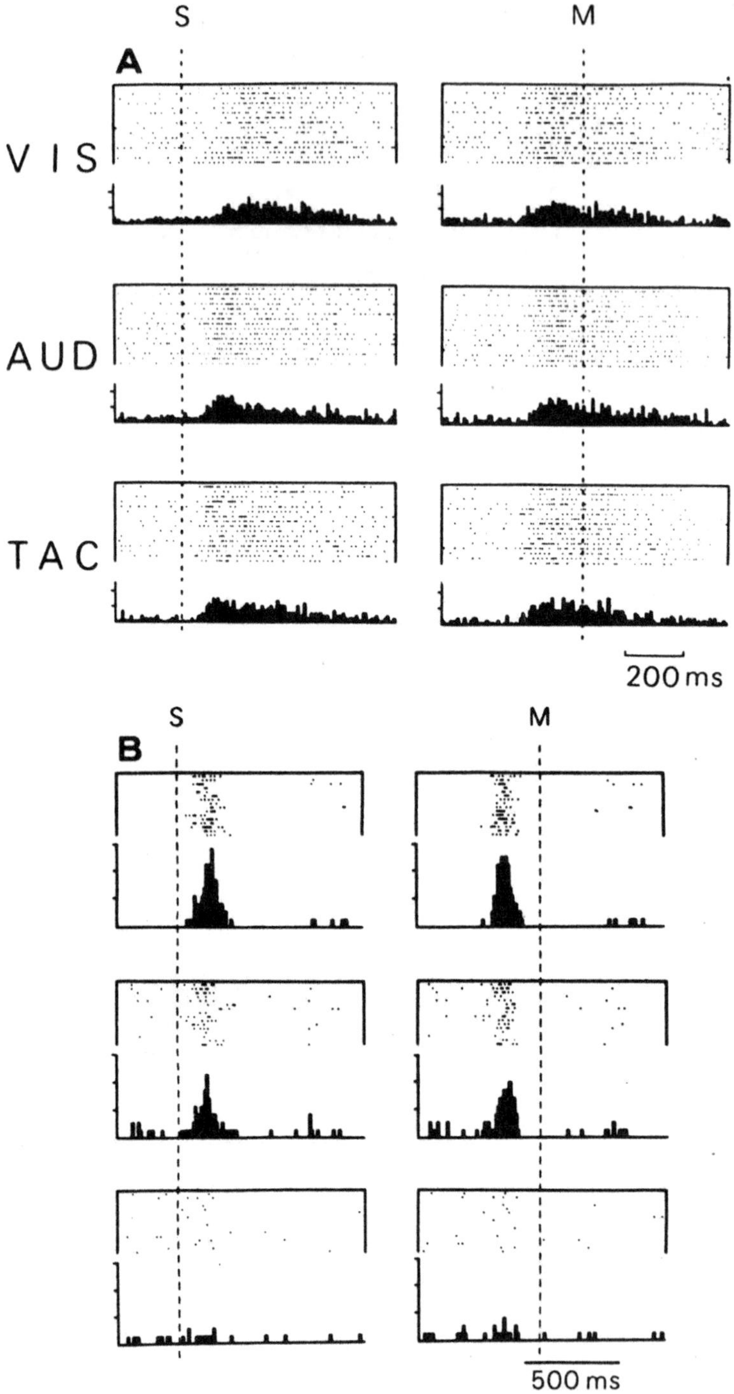

Fig. 13

In a previous study on anaesthetized monkeys, we have described results which also provided suggestive evidence for a 'setting' role of the SMA. Conditioning electrical stimulation was found to inhibit the responsiveness of area 4 neurones to controlled stretch stimuli of a forearm muscle (*Wiesendanger* et al. 1975). We have recently extended these experiments on awake monkeys subjected to load perturbations (*Hummelsheim* and *Wiesendanger*, unpublished). Again, some of the neurones tested showed a diminished responsiveness when the load perturbation was preceded by conditioning electrical stimulation of the SMA (Fig. 12).

In a study with the same experimental paradigm, *Tanji* and *Kurata* (1982) compared SMA neurones and neurones of the precentral motor cortex. The task of the monkey was to perform, or not to perform, a simple wrist movement, depending on a sensory cue signal (visual, auditory, or tactile). In general, it was found that SMA cells were more closely related to the cue signal whereas precentral cells were best related to movement onset. The latency of the response to the cue signal (except for the tactile signal) was shorter for SMA neurones, and the onset of the response was usually brisk. That the response of SMA neurones was not simply a sensory response was reflected in the fact that it occurred only when the cue signal was also followed by a motor response. Some neurones responded similarly to all three modalities tested; others responded differently to each modality (Fig. 13).

'Set' and 'instruction'-related cells are common in the primate's frontal cortex and have been found in abundance in the prefrontal association cortex (see e.g. *Kubota* et al. 1974). It was suggested by *Kubota* (1985) that there is a gradual shift in the occurrence of such cells, which are most frequent in the prefrontal cortex and least frequent in the motor cortex, and that lateral and medial area 6 may be in an intermediate position. It is clear that response patterns similar to those observed by *Tanji* and his associates can also be found in the motor cortex, i.e. that the SMA is not unique in this respect. However, the incidence of setting-related neurones was relatively low in the motor cortex (*Tanji* and *Evarts* 1976).

Fig. 13A,B. Responses of SMA neurones to sensory cue signals. Time histograms and dot displays of two SMA neurones were recorded in a monkey trained to perform wrist flexions and extensions in response to visual (*VIS*), auditory (*AUD*), and tactile (*TAC*) stimuli. The activities were aligned either by stimulation onset (*S*) or by movement onset (*M*). Note that unit **A** increased its firing rate in a similar fashion for all three signals, and each time the activity was also time-locked with the movement. The **B** neurone correlated with the movements triggered by visual and auditory cue signals, but not with those triggered by the tactile cue. The ordinate scale unit is 40 impulses per second in **A**, and 28 impulses per second in **B**. (*Tanji* and *Kurata* 1982)

3.3 Other Correlations of Cellular Activity in the SMA

Present knowledge of the functional properties of SMA neurones and their role in motor performance is still very rudimentary. For instance, there is little information about possible correlations with *movement parameters*. *Smith* (1979) found no relation between discharge frequency and rate-of-force change when the monkeys had to squeeze a wafer in order to obtain an alimentary reinforcement. The majority of SMA cells displayed an increased static firing rate during the time the monkeys were pinching the manipulandum. The author suggested that the variability in the timing of the increased discharge rate was indicative of a difference in recruitment thresholds with respect to the generated force output (these SMA cells behaved similarly to many precentral cells).

There are also few investigations on the direction specificity of SMA cells (cf. *Kalaska* et al. 1983) or their relation to the speed of a movement (cf. *Hamada* 1981). *Tanji* and *Kurata* (1982) reported that about 50% of SMA neurones tested in a hand movement task were direction-specific. *Brinkman* and *Porter* (1983) noted, however, that SMA cells were sometimes "... not related to specific movement, but rather showed changes in discharge frequency throughout a movement sequence".

The topological relationships of the SMA have already been discussed in Sect. 2.3. A rostrocaudal body scheme seemed to be confirmed in the single-unit studies of *Brinkman* and *Porter* (1979), *Wise* and *Tanji* (1981), and *Tanji* and *Kurata* (1982). However, it is our impression (detailed in Sect. 2.3) that the topological relationships of the SMA are much more complex. The difficulties in drawing conclusions about topology from behavioural studies are manifold. Of particular importance is the possibility that even in a distal motor task subtle changes in proximal and trunk muscles are likely to occur, but are difficult to control. It is significant that even in the precentral cortex a considerable proportion of neurones were found to be related with more than one task. Such neurones may reflect the collateralization of descending neurones (cf. also *Matsunami* and *Hamada* 1983).

Another point which deserves further study is whether SMA neurones are implicated in *bilateral control* of the limbs, as particularly advocated by *Brinkman* and *Porter* (1979, 1983). *Tanji* and *Kurata* (1981) reinvestigated this problem and found that only less than 10% of the neurones were equally well correlated with left and right hand movements. This is a smaller percentage than found for cells in area 4 (*Matsunami* and *Hamada* (1981). It is conceivable that many of the 'bilateral' SMA cells of *Brinkman* and *Porter* (1979) may have been best related to trunk or neck muscles, which were likely to be implicated in the complex movement task. It is noteworthy in this respect that *Tanji* et al. (1980) briefly described 26 SMA neurones which were best related to the activation of paravertebral muscles.

Finally, *Brinkman* and *Porter* (1979) discovered a small sample of neurones in the rostral SMA which was thought to be implicated in *visual experience.* No formal testing of the visual stimulus and of ocular movements was done, however. The conclusion was based on increased cell firing observed while a food-retrieving movement was performed to one side, but not to the other side of the body. This association was conditional on visual feedback. Saccade-related unitary activity was recently found in the dorsal portion of medial area 6. From this same region saccades could be elicited by microstimulation (20–50 μA) at latencies of 45–90 msec (*Schlag* and *Schlag-Rey* 1985). The possible role of the SMA in visuomotor control and visual attention deserves further experiment, also in view of the striking anatomical relationship of the rostral SMA to the frontal eye field.

4 The Role of the SMA in Motor Control as Revealed by Lesion Studies in Monkeys

4.1 General Behavioural Effects and Short-Term Deficits

All investigators seem to agree that SMA lesions in monkeys do not manifest themselves in gross changes of motor behaviour. Locomotion, climbing, feeding, and grooming were reported to be normal. It was noted that sometimes the arm ipsilateral to the lesion was preferred for grasping and that there was some neglect of the contralateral arm (*Brinkman* 1984), but this was seen only for a few weeks after lesioning.

During the early postoperative period, *Brinkman* (1984) noted also some deficits in a formal testing situation requiring the use of independent finger movements. Food morsels were retrieved, by means of a precision grip, from slots oriented at different angles over the surface of a board. After undergoing a unilateral SMA lesion, the monkey was still able to retrieve the food morsels from the slots, but a frame-to-frame analysis of video recordings revealed a number of subtle changes in the performance. Before the lesion, the hand was adequately 'shaped' during the reaching phase for grabbing the food morsel; after surgery, the pregrasp shaping of the hand was lacking. The grasp was somewhat jerky and less smooth than before the lesion. Whereas the normal monkeys had a stable course when they proceeded from one slot to the next, the lesioned monkeys had an erratic course. Interestingly, the unilateral lesion appeared to affect both sides.

In a recent abstract, *Halsband* (1983) reported that animals with SMA lesions had "severe difficulty in the ability to sequence movements".

4.2 Possible Effects on Muscle Tone

The question which cortical area is responsible for the development of spasticity has been debated for a long time. It appears now that prominent spasticity is rarely seen in monkeys unless motor-cortical lesions are extensive and bilateral (*Tasker* et al. 1975). In early investigations, made without quantitative measuring of muscle tone, it was believed that the premotor cortex of lateral area 6 (*Fulton* 1949), the 'strip' region of area 4 (*Hines* 1936), or the SMA (*Travis* 1955; *Travis* and *Woolsey* 1956; *Pinto-Hamuy* 1956) were the crucial areas. However, the results of *Coxe* and *Landau* (1965) obtained in SMA-lesioned monkeys were rather disappointing in this respect: "slight, but inconsistent increase in tonus was noted on passive manipulation of joints, and particularly upon elevation and rotation of the shoulder (occasionally, increase in biceps and quadriceps tone was found)." *Denny-Brown* (1967) reported that monkeys with bilateral SMA lesions developed "a mild flexion rigidity". *Brinkman* (1974) described monkeys with extensive lesions of the motor cortex which included the SMA; they developed a flexion posture with adduction in the shoulder and flexion in the elbow and in the wrist. When the SMA was spared but the monkeys were subjected to otherwise similar motor cortex lesions, the postural changes did not occur. Muscle tone was not measured in any of the above investigations. As long as quantitative assessments of stiffness and stretch reflexes in animals with SMA lesions are lacking, it is not possible to draw any conclusion about the role of the SMA in controlling muscle tone.

4.3 The Role of the SMA in the Automatic Grasp Response

Automatic grasp responses to tactile stimuli are well known to occur in patients with frontal lobe lesions (*Adie* and *Critchley* 1927). The 'instinctive grasp reaction' (*Denny-Brown* 1966) and 'forced grasping' (*Fulton* 1949) have been interpreted as release phenomena occurring as a consequence of premotor cortex lesions in monkeys (for a discussion of terminology cf. *Wiesendanger* 1981a). Later, *Penfield* and *Welch* (1951) and *Travis* (1955) reported that the critical site for the development of forced grasping was the SMA; in Penfield's words. "... forced grasping is a specific sign of removal of the supplementary motor area in the monkey" (*Penfield* and *Jasper* 1954). The problem has been reinvestigated recently in monkeys trained to perform a conditioned grasp response for obtaining an alimentary reinforcement (*Smith* et al. 1981). The task consisted of squeezing a small wafer with the thumb and another finger and exerting a given pressure for a 1-s period. The authors describe the deficits as follows:

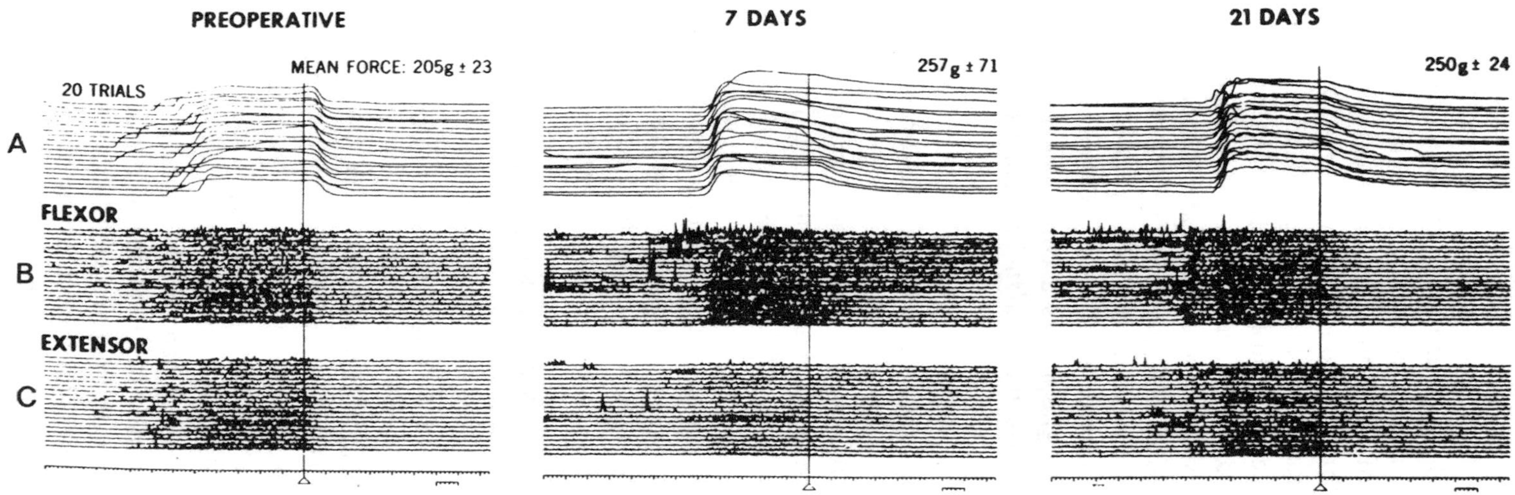

Fig. 14A–C. Recordings during performance of a conditioned finger pinch by a monkey before and 7 and 21 days after sustaining a contralateral SMA lesion. **A** Position traces of 20 consecutive trials at each time; **B, C** records of the corresponding rectified EMG signals for flexor (**B**) and extensor (**C**). The monkey was trained to release the pinch in response to a signal, which is marked by the *vertical lines*. Note delay of the release movement, which is especially pronounced in the early postoperative period. (*Smith* et al. 1981)

"Within 24 hours after the SMA ablations, forced grasping appeared in the contralateral hand and persisted for one to three weeks. Grasping was easily elicited by cutaneous stimulation of the volar surface of the hand and could be augmented by stretch of the finger flexors. The most striking disturbance of the conditioned precision grip was the animal's inability to release the force transducer from its grasp with the contralateral hand. In contrast, release from the ipsilateral hand remained normal." These findings were taken to indicate that the SMA exerts an inhibitory control on the motor cortex (cf. also Sect. 3.2). Electromyograms of flexor and extensor activities of the contralateral hand are shown in Fig. 14. Alternatively, corticospinal neurones of the SMA "... may act directly to suspend the effects of somatic segmental reflexes allowing voluntary movements" (*Smith* et al. 1981). Other investigators, however, failed to observe forced grasping as a consequence of SMA lesions (*De Vito* and *Smith* 1959; *Coxe* and *Landau* 1965; *Brinkman* 1984).

4.4 Reaction Time Studies in SMA-Lesioned Animals

The delayed release of a conditioned grasp response described in the preceding paragraph could also be interpreted in terms of prolonged reaction times (*Smith* et al. 1981). Surprisingly, there is no other study in which the generation of a rapid limb movement was tested in a reaction time situation. If the SMA is pivotal in the initiation of movements, one could expect an increase in reaction times following bilateral SMA lesions. On the other hand, triggered movements may be unaffected by SMA lesions if the contention of *Libet* et al. (1982, 1983a) is correct that the SMA is chiefly involved in the generation of 'free-willed', self-initiated movements. This question clearly calls for further study.

4.5 The Role of the SMA in Bimanual Coordination

In two monkeys with bilateral SMA lesions, *Brinkman* (1981, 1984) observed a peculiar deficit in bimanual coordination. The task, previously used by *Mark* and *Sperry* (1968) in testing animals with callosal sections, consisted in retrieving currants wedged into holes of a Perspex plate. The normal animals immediately learned to push out the currant with the index finger of one hand while cupping the other hand underneath to catch the currant. After lesioning of the SMA, both hands were used in a mirror fashion with the two index fingers pushing against each other from above and below. This approach was of course not very successful; however, one of the animals, whose lesion was opposite to the nonpreferred hand, gradually "... developed a number of 'tricks' to mask the deficits". The interpretation of the results was as follows: the SMA, through its strong connections with the motor cortex, transmits instructions for a motor sequence to the motor cortex. Normally, if two different tasks are required of the two hands, each side influences the ipsilateral motor cortex. After one-sided lesions, the unimpaired SMA would transmit the 'programmes' for bimanual tasks to both sides. Interruption of the corpus callosum would then, according to this hypothesis, prevent the second hand from doing the 'same' thing as the hand ipsilateral to the intact SMA. *Brinkman* (1984) found that section of the corpus callosum indeed immediately abolished the bimanual deficit. When monkeys were subjected to *bilateral* SMA lesions, the bimanual deficit also failed to occur (*Brinkman* 1983).

This experiment is of great interest, but other interpretations are at least conceivable. Also, a number of questions arise from this experiment for which we have no ready answers. Thus, the 'programme' issued by the SMA would be of a higher order since the 'mirror movements' after a unilateral SMA lesion require different sets of muscles, and the spatio-temporal sequence of motor-cortical excitation must be different on either side. A second problem is that apparently centres of the brain other than the SMA were capable of 'programming' the motor cortex, since correct bimanual performance was achieved immediately after corpus callosum section and bilateral SMA lesions. The deficit seen with a unilateral SMA lesion and intact corpus callosum thus appears to produce an imbalance in the programming of the motor centres of the two hemispheres (*Brinkman* 1983).

The neural mechanisms of bimanual control are of great interest and obviously play an important role in the execution of motor skills. Yet, one has to admit that virtually nothing is known about these mechanisms. The lesion approach, as reported above and in previous work on callosal lesions (*Mark* and *Sperry* 1968), is an encouraging first step in unravelling the complexities of involved mechanisms.

4.6 Effects of SMA Lesions on Vocalization

In view of the old observations that in man the SMA is implicated in vocalization (cf. Sect. 5.2), it was interesting to investigate whether a similar role is true for subhuman primates as well. *Kirzinger* and *Jürgens* (1982) reported that bilateral lesions of the SMA in squirrel monkeys indeed interfered with vocalization. The frequency of spontaneous overall vocalization diminished significantly, and this was found to be due essentially to the decrease of a single type of vocalization, namely the so-called isolation peep. This is a self-initiated long-distance call which is not triggered by other calls. It is presumably a learned and not a genetically preprogrammed utterance (*Kirzinger* and *Jürgens* 1982). This deficit was specifically related to the anterior portion of the SMA. The differential effect on only one type of vocalization is significant because it proves that the deficit is not a simple motor disorder of phonation. More recently, *Sutton* et al. (1985) found that vocal response latency was greatly increased in a monkey (M. mulatta) with a bilateral anterior SMA lesion, but also in a monkey with a more anterior lesion in medial areas 8B and 9.

The above observations are of considerable interest because they are so far perhaps the clearest results from animal work that seem to support the contention based on human studies that the SMA has a 'higher' programming role in self-initiated motor behaviour.

5 Evidence for the Implication of the SMA in Voluntary Movements in Man

5.1 Lesions Involving the SMA of the Human Brain

The position of the SMA in the human brain is illustrated in Fig. 15. The case reports are of patients who underwent an excision of the SMA because of an epilectic focus or of a tumour, or who suffered an occlusion of the anterior cerebral artery. It must be realized that in many cases either the exact boundary of the lesion was not known or the lesion proved to encroach also on neighbouring cortical areas. The first case descriptions came from the Montreal Neurological Institute (*Penfield* and *Welch* 1951; *Penfield* and *Jasper* 1954). Most case histories relate to unilateral lesions; I have knowledge of only one brief report describing bilateral lesions in two cases of parasagittal meningioma (*Woolsey* et al. 1979).

The most striking symptoms of the Montreal cases were a *slowing of movements* on the side contralateral to the lesion (noted e.g. as slowing in typewriting) and *forced grasping.* Typically, these symptoms gradually diminished, but it was noted that some "... slowness when repeated or alternating movement is attempted on the opposite side of the body" was present even as long as 1 year after the lesion. Forced grasping was a prominent sign for 1–2 weeks after the lesion; during this time the patients had to exert considerable effort to open their hands. A patient was described who "... amused herself for a week or two by placing things in the affected hand and watching its involuntary grasp movement". These findings have been confirmed in subsequent reports by *Rubens* (1975), *Woolsey* et al. (1979), *Goldberg* et al. (1981), and *Gelmers* (1983).

Talairach and his colleagues (*Laplane* et al. 1977) described a contralateral reduction of voluntary movements, termed *akinesia,* in three

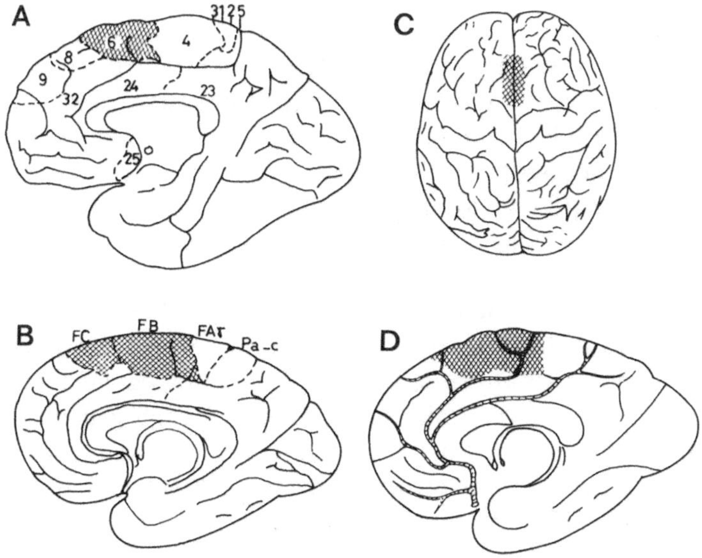

Fig. 15A–D. The SMA of the human brain. **A, B** Cytoarchitectonic areas according to the maps of **A**, *Brodman* (1925) and **B** *von Economo* and *Koskinas* (1925). According to cytoarchitectonic definition, the *cross-hatched areas* (medial area 6 in **A** FB-FC in **B**) correspond to the SMA. **C** Dorsal view with approximate location of the SMA. **D** The SMA in relation to the territory supplied by the anterior cerebral artery (*Bailey* 1933)

patients who underwent a unilateral excision of the SMA because of an intractable epileptic focus. The patients' faces were reported to be almost motionless and to exhibit "an emotional type facial palsy" (cf. also *Laplane* et al. 1976). Furthermore, the patients also had difficulty in gazing sidewards. Alternating and serial hand movements were performed awkwardly. Spontaneous speech was markedly reduced. Whereas most symptoms gradually subsided, the poverty of movements and of spontaneous speech remained for a long period. In an abstract, *Damasio* and *Van Hoesen* (1980) mention four patients who had localized lesions in the SMA (as evidenced by computed tomography [CT] scans) and suffered from a general lack of "drive for willed movements, including speech and gesture".

In two patients with a medial frontal cortex infarction (which also included some of the leg representation in the paracentral lobule), *Goldberg* et al. (1981) observed a peculiar manipulation disorder which they termed *the alien hand sign.* One of the patients would reach out with the right arm for many objects, apparently unable to inhibit this behaviour except by holding the affected arm with "the more obedient arm". Furthermore, "... the right arm was noted to interfere with tasks being performed by the left arm". The other patient had a similar tendency to produce compulsory goal-directed movements which, according to the patient, were not initiated by herself ("the arm will not do what I want it to do"). *Goldberg* et al. (1981) assumed that these deficits, including the lack of speech initiation and the strong grasp reflex, were caused essentially by a lesion of the SMA, but the authors did not exclude participation of neighbouring structures.

Spasticity was not noted in the above cases. However, *Russel* and *Young* (1969) described a series of patients who suffered from missile wounds in the region of the paracentral lobule on the medial surface. The lesions included the foot representation of area 4, but may have extended forward to include some of the SMA. Interestingly, lesions with this localization led to an *immediate* spasticity in both legs and sometimes also in one arm. Whether the SMA played a role in the generation of spasticity in these cases is unclear. With the present improved possibilities of localizing neurological lesions rather pecisely in patients, the old but still unresolved problem of localization of spastic states ought to be reinvestigated.

5.2 The Role of the SMA in Vocalization and Speech

Arrest of vocalization and speech are by far the most frequent effects produced by electrical stimulation of the SMA in neurological patients. In the Paris cases, speech arrest was observed in 50% of all incidences of intracortical stimulation (*Bancaud* et al. 1985; cf. also reports by *Talairach* and *Bancaud* 1966 and *Chauvel* 1976). A typical sequence, upon prolonged electrical stimulation, was acceleration of loud counting – arrest of counting – vocalization. However, vocalization may also occur as the first manifestation during electrical stimulation. The motor effects produced by SMA stimulation (for review of older literature cf. *Wiesendanger* 1981a) occurred less frequently than vocalization and speech arrest: arm elevation and head turning in 33%; deviation of the eyes in 17%; leg movements in 6% of the cases. The authors assumed that vocalization and speech arrest were not due to spread of current to the cingulate cortex because electrical stimulation of area 24 of the anterior cingulate cortex evoked characteristic distal hand and finger movements but not vocalization (*Bancaud* et al. 1976).

The motor manifestations of electrical stimulation have a striking resemblance to epileptic seizures with a focus in the SMA, with their contraversive movements of the trunk, elevation of arms, speech arrest, and saccadic vocalization (see e.g. *Talairach* and *Bancaud* 1966; *Woolsey* et al. 1979; *van Buren* and *Fedio* 1976). Evidence is now accumulating that lesions in the region of the SMA may greatly interfere with language. Cases of speech disturbances associated with parasagittal frontal lesions were reported by *Chusid* et al. (1954), *Guidetti* (1957), *Környey* (1975), *Rubens* (1975), *Alexander* and *Schmidt* (1980), *Damasio* and *Van Hoesen* (1980), *Jonas* (1981), and *Goldberg* et al. (1981). Each of the three patients of *Laplane* et al. (1977), referred to above, had a conspicuous poverty of verbal expression. In the case description of *Masdeu* et al. (1978), it was specified that the lesion was on the dominant side and that the concomitant difficulty in writing persisted until the death of the patient. In two patients with bilateral lesions of the SMA, described by *Woolsey* et al. (1979), "profound aphasia" was noted.

However, it is possible that language is also affected by an SMA lesion on the non-dominant hemisphere, as recently reported by *Gelmers* (1983). The patient "did not initiate conversation" and "spoke very haltingly in reply to questions", but was quite capable of repeating complex phrases (this language disorder has also been termed 'transcortical motor aphasia').

In all cases, a considerable, but not total recovery took place. *Green* et al. (1980) described a patient in whom a surgical excision of the dominant SMA was necessary to remove a large porencephalic cyst which caused "partial seizures of the postural type". Postoperatively, the deficits were minimal, with a transitory slowness of movements but without speech disorder or forced grasping. Perhaps recovery had already occurred in this long-standing lesion, which was present before surgery.

In summary, the observations in patiens with SMA lesions and the reported effects of electrical stimulation in the human brain tend to confirm the original concept of *Penfield* (*Penfield* and *Jasper* 1954) that the SMA has to be considered a speech area. Some uncertainties remain, however, because in many cases the exact extent of the lesion was not known or proved to involve other cortical areas as well. In particular, inclusion of the cingulate cortex may well have contributed to the reported language disorders.

5.3 A New Outlook on the Function of the SMA Derived from Measurements of Changes in Regional Cerebral Blood Flow

The method of measuring regional cerebral blood flow (rCBF), developed by *Lassen* and *Ingvar* (1972), brought a new dimension to the exploration of cortical functions in the living human brain. The method rests on the reasonable assumption that increases in cerebral blood flow are associated with increased metabolism and thus with increased neuronal (as well as glial) activity. It was shown that sensory stimulations 'activated' the appropriate cortical areas, and that reading and speaking produced increases of rCBF in the appropriate language areas of the dominant hemisphere (*Ingvar* and *Schwartz* 1974). What, then, was the outcome when patients

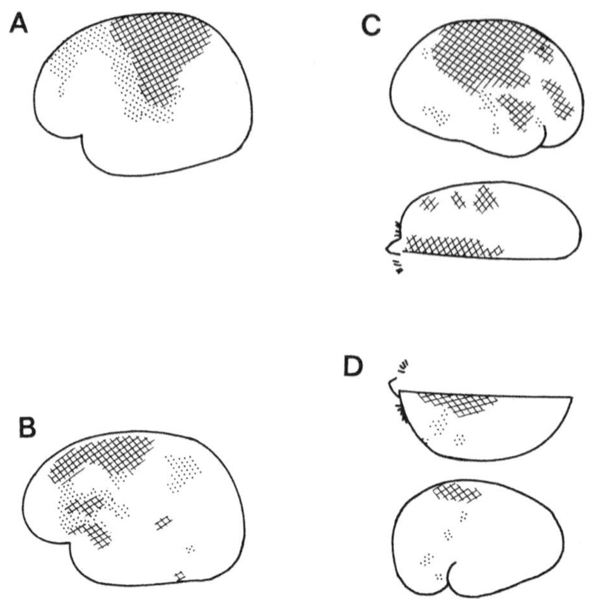

Fig. 16A–D. Changes in regional cerebral blood flow (rCBF) occurring in the human brain as a consequence of performance or mentation of various motor tasks. Increases in rCBF of more than 20% are shown as *cross-hatched* areas; below 20%, as *punctate areas*. **A** Increase of rCBF (relative to rest) in the pre- and postrolandic areas when contralateral hand movements were performed. **B** Shift of the increased rCBF towards frontomedial regions when hand movements were conceived but not actually performed. Note that the changes occurred over a wide fronal territory. **C** A complicated sequence of finger movements, performed with the left hand, resulted in a widely distributed increase of rCBF in the frontoparietal lobe. *Top,* lateral views; *bottom,* dorsal view, obtained with the same test in another patient. **D** The same sequence of finger movements was mentally conceived but not performed. Dorsal view (*top*) and lateral view (*bottom*) are from two different patients. Note that the increase in rCBF was concentrated in the medial-frontal cortex. **A** and **B** adapted from colour Fig. 2 of *Ingvar* and *Philipson* (1977); **C** and **D**, from colour Fig. 2 of *Roland* et al. (1980a)

were asked to perform motor tasks? As expected, simple, repetitive movements, such as clenching the fist, were shown to activate the somatotopically appropriate region of the contralateral Rolandic cortex (*Olesen* 1971; *Foit* et al. 1980). Interesting were the observations that the *postcentral* area also displayed a surprisingly high change of rCBF (see e.g. *Ingvar* 1977) and that hand movements resulted in activation of a medial frontal focus, which persisted even when the movement was not actually performed but only conceived (Fig. 16A and B; *Ingvar* and *Philipson* 1977). This observation then led to the new idea that the medial frontal cortex may play a major role in the programming of movements. This was expressed in the influential paper by *Roland* et al. (1980a), in which a thorough investigation of rCBF changes accompanying hand movements and movement mentation was reported. The medial frontal focus was now narrowed to the SMA, which was activated bilaterally and only if the motor task was not a trivial one. A trivial, repetitive task or steady contractions had little effect on the rCBF of the SMA (*Roland* et al. 1980; *Orgogozo* and *Larsen* 1979). The most intriguing observation was made when patients who had practised a complex task (thumb-to-fingers in a complicated sequence) were asked to imagine the sequence without actually performing it. This mental act manifested itself in a selective activation of the SMA on both sides (Fig. 16C and D). SMA activation was therefore a brain correlate of the translation process from the 'idea' to the actual performance of the movement. The results also suggested that the SMA is involved only if the voluntary movements demand conscious attention, not if they are performed automatically. This is the interpretation given by the authors, which then led *Eccles* (1982) to formulate his hypothesis about the role of the SMA in movement initiation as mentioned in the Introduction.

The importance of rCBF studies on the human brain during the planning and execution of various motor tasks is undeniable; the method has already led to increased efforts towards understanding the SMA, which appears to be particularly involved in these processes. In discussing this work, it is, however, also necessary to be aware of the limitations of the method. In the following, I will point out some of the problems:

(1) Patient Selection. Thus far, rCBF measurements had to be performed with invasive methods, and the available results are from neurological patients with a clear diagnostic indication for the use of the method. This situation may change rapidly since non-invasive methods, such as positron emission tomography, have been introduced, which will permit the investigation of rCBF in normal subjects.

(2) Temporal Resolution. Essentially, the method provides indirect information about changes in the total metabolism of the brain surface occurring about 40 s before the measurements are taken. It is during this time that the behavioural tests have to be performed. The percentage of deviation from control measurements taken at rest may be interpreted as causally related to the behavioural performance. The lack of temporal resolution makes it practically impossible to say in what temporal sequence the metabolic change of a given brain region occurred with respect to the task. It is, thus, difficult to establish whether an observed brain process is the cause or the consequence of the behavioural act.

(3) Spatial Resolution. The method points appropriately to the *focus* of changes in rCBF, but it is difficult to establish the *extent* of altered metabolism. The changes are conventionally mapped by means of a colour

code indicating the percentages of changes from rest values. We do not know the lower limit of the percentages of significant changes. Thus, in the earlier paper by *Ingvar* and *Philipson* (1977), which gives a comparison of motor performance and 'motor ideation', the medial frontal focus was large and exceeded the SMA, and a significant increase in rCBF was also noted in the temporal lobe. More generally speaking, the method has its evident limitation in correlating the brain area showing relative changes in its metabolism with areas having a distinct cytoarchitectonic and hodological definition. In particular, uncertainties remain about the possible implication of neighbouring cortical areas of the SMA such as the premotor cortex on the lateral surface, the medial prefrontal cortex, and the underlying cingulate cortex.

(4) Specificity of the SMA Involvement. It is not unequivocally established that the SMA is specifically involved in movement mentation. Voluntary saccades were found to strongly activate the SMA (*Fox* et al. 1985). It should be noted that changes of rCBF in the region of the SMA were also noted when peripheral nerves were stimulated electrically beyond the twitch threshold or when the subjects imitated the thumb twitches (i.e. a simple movement) by voluntary effort (*Foit* et al. 1980).

To summarize, it appears that the SMA of the human brain is indeed a 'hot spot' with respect to movement mentation. However, future studies will have to clarify the temporal relation of the SMA activation to the behavioural act, and the precise extent of the implicated brain region.

New methods have recently been developed which provide a much better spatial as well as temporal resolution. Computerized positron-emission tomography is also applicable to deep structures and may be ethically used with normal subjects (*Roland* et al. 1982; *Fox* et al. 1985). More specific parameters such as glucose consumption or transmitter release may be assessed. Future studies certainly will provide answers about the correctness of the interpretations derived from classical rCBF studies.

5.4 Movement-Related Slow Potentials Recorded From the Vertex in Man

'Readiness potentials' (RPs) occurring about 1 s before onset of a self-paced movement were first described by *Kornhuber* and *Deecke* (1965) and *Gilden* et al. (1966) and have been taken to indicate readiness to move. The RPs have a maximum over the vertex (*Deecke* et al. 1969), and *Deecke* and *Kornhuber* (1978), in a short communication, proposed that the RP over the vertex signals participation of the SMA. In parkinsonian patients, it was found that the RP over the Rolandic cortex may be strongly depressed whereas it remains prominent over the vertex. The authors therefore concluded that the RP generated by the SMA is independent from that of the Rolandic cortex. Since RPs can be recorded bilaterally over wide areas of the cerebral cortex, the special role of the SMA in the generation of these brain events is not obvious. The complicated geometry of the mesial cortex makes it very difficult to deduce the importance of a cortical area in generating the RP on the basis of the amplitude of the potential. In the monkey, *Sasaki* and co-workers were able to record premovement potentials in triggered and self-paced movements which were similar to the RP. In a recent paper, *Gemba* and *Sasaki* (1984) reported, however, that the premovement potential recorded intracortically in the SMA during the performance of self-paced movements was of smaller amplitude. With visually triggered movements, the premovement potential in the SMA occurred later than in the motor cortex. *Deecke* and *Kornhuber* (1978) and *Boschert* et al. (1983) emphasized the particular importance of the SMA in generating the slow potentials associated with the preparatory phase of a volitional movement. However, one cannot escape the conclusion that the distribution of the premovement potentials is rather wide, and that it probably also, but not exclusively, covers the SMA.

There is now a large amount of published data on various aspects of premovement potentials, but since I have no particular competence to deal with this subject, I will refer the interested reader to a number of recent reports. *Libet* and co-workers (1982, 1983a,b) were particularly interested in studying 'freely voluntary movements' and in obtaining estimates of the temporal relations between the subjective awareness, of the urge to move and the onset of the RP recorded at the vertex. The work of *Grünewald* and *Grünewald-Zuberbier* (1978, 1983) is concerned with premovement potentials and larger potentials following these during aimed movements towards a target (aiming potentials). *Jung* et al. (1982) described the slow potentials associated with linguistic and writing tasks. Finally, *Lang* et al. (1983) have been interested in the slow potentials associated with visuomotor learning tasks. For further related work which cannot be dealt with in this review, see also the papers by *Deecke* et al. (1983), *Boschert* et al. (1983), *Kutas* and *Donchin* (1980), *Shibasaki* et al. (1980), and the review by *Haider* et al. (1981).

6 Synthesis on the Function of the SMA

6.1 'Low-Level' Controls of the SMA

Many results of monkey experiments seem to indicate that the SMA exerts motor control functions at a relatively low level. Thus, the SMA was found to be organized in discrete efferent zones with a fairly direct control on spinal motoneurones. The evidence for this rests on the existence of corticospinal neurones, on positive intracortical microstimulation effects, and on the observations of stimulus-triggered facilitation of ongoing EMG activity. In the last-named situation, a *single,* short pulse of less than 30 μA is likely to activate only a small volume of cortical tissue (cf. *Cheney* and *Fetz* 1985). As in the motor cortex, a contingent of SMA neurones was found to respond to somatosensory signals via relatively direct pathways. Lesion experiments have sometimes revealed changes in posture and muscle tone, and a release of tactile grasping. Finally, single-unit recordings in the awake, performing monkey revealed patterns of neuronal activity that covaried with learned movements much in the same way as previously described for motor cortex neurones. The problem of timing has so far not been resolved; in the experiments of *Brinkman* and *Porter* (1979), the onset of cell firing in the SMA appeared earlier with respect to movement onset than found in similar experiments on the motor cortex. But *Smith* (1979) found just the opposite to be true for a manual task.

Considering the above-mentioned experimental results, it appears to me that they are not *principally* different from those obtained in the motor cortex. It is likely, however, that there are *quantitative* differences: there are considerably fewer corticospinal cells in the SMA (and they are all of small calibre) than in the motor cortex; there is more overlap in the somatotopical organization; the effects of intracortical-train and single-pulse stimulation are much less conspicuous than in the motor cortex; and finally, the sensitivity to somatosensory signals is restricted to discrete subsets of neurones of the SMA (mainly non-projection neurones).

It is also important to note that the above results were mainly seen in the *posterior half* of the SMA, which appeared to be mostly concerned with the control of shoulders and trunk.

6.2 'High-Level' Controls of the SMA

Let us first consider the evidence derived from animal experiments. One important outcome in the experiments of *Tanji* and *Kurata* (1983, cf. Sect. 3.2) was the finding that neuronal activity was often related to *instruction signals* rather than to movement execution. Such neurones may be considered to subserve *programming functions.* Also, the neuronal response appeared to be more tightly linked with the sensory cue signal used to trigger the movement than with the movement per se. At the behavioural level, lesion experiments of *Brinkman* (1981, 1984) revealed three types of deficits which can be attributed to a loss of high-level controls: (1) a striking deficit of bimanual coordination after unilateral SMA lesions; (2) the loss, in a food retrieval task, of the preoperative straightforward strategy of removing food morsels from one slot after another; (3) the observation that the lesioned monkeys failed to shape their hand in anticipation of the impending grasp. Some of these deficits were discrete, requiring a close frame-by-frame analysis of video tapes, and they improved rather quickly with time. Finally, the important discovery of *Kirzinger* and *Jürgens* (1982) must be mentioned, that a learned vocalization, the 'isolation peep', was abolished following bilateral lesions of the anterior portion of the SMA of squirrel monkeys. Taken together, these findings show that in subhuman primates the SMA does indeed play some role in the initiation of learned movements. In contrast to human cases, it appears, however, that the general motor behaviour and motor 'drive' in subhuman primates are hardly affected by SMA lesions.

In *man,* the evidence for such a 'higher' role of the SMA is more impressive. To summarize briefly: lesions in the region of the SMA lead to a general reduction of voluntary movements, a kind of akinesia, and to a striking poverty of *spontaneous* speech. Readiness potentials, which precede voluntary movements by almost 1 s, have been found to be largest over the SMA. Changes in rCBF in the SMA occur during the preformance of non-trivial motor tasks and even during movement mentation. These are the main arguments advanced in favour of SMA involvement in the preparatory phase of voluntary movements.

It is tempting to assume that the structural relationships to the basal ganglia, the cingulate cortex, and other association areas may be very important to these high-order functions of the SMA.

In my own work, I have not addressed the question whether the SMA participates in such high-order controls, and I have therefore not gone into a detailed critical discussion of such studies, especially not of those on the human brain. It appears that some of the observations made in patients may not exclusively be attributed to the SMA. Future studies, performed with methods providing better temporal and spatial resolution, will presumably soon bring the answers. Lastly, it should be realized that complex motor deficits, such as poverty of movements, the 'alien hand syndrome', and poverty of spontaneous speeck ('transcortical aphasia') have sometimes been observed with lesions in the frontal lobe that were outside the SMA.

6.3 A Proposition that the SMA Exerts a Double Control: at a 'Low' Spinal Level and at a 'High' Integrative Level

Given the present evidence, which it is hoped, will be strengthened by future studies, it is suggested that the SMA exerts a double control: at a relatively low level and at a hierarchically higher level, i.e. in initiation processes of voluntary movements. It may now be appropriate to ask what the significance of this proposed double role might be. At first sight, the two functional levels do not fit together at all, and it may even be conjectured that the low-level control of the SMA is a functional attribute left to the brain of subhuman primates, and that in man, the high-level control largely dominates and replaces the low-level controls seen in monkey studies. To some extent, this conjecture may be justified. When comparing the simian and the human brain, the 'intermediate' cortex of *Campbell* (1905), which includes lateral and medial area 6, is relatively more expanded than area 4 in the human brain. It may therefore well be that the human SMA subserves new, 'higher' functions. However, this does not mean that the 'low-level' controls of the SMA were 'lost' in the human brain. Thus, the effects of lesions (changes of muscle tone, forced grasping) and of electrical stimulation indeed suggest that the SMA of the human brain exerts more direct influences on the spinal cord.

One might reconcile the two seemingly opposing views on the SMA function by speculating that *the direct influence of the SMA on the motor apparatus sets the appropriate excitability level of spinal circuits, allowing anticipatory postural adjustments in advance of voluntary movements.* A similar idea has been proposed by *Humphrey* (1979). That postural preparation and adjustments accompany most voluntary movements was clearly recognized by *Hess* (1943) and has again been emphasized in a number of more recent studies (e.g. *Gurfinkel* et al. 1971; *Bouisset* and *Zattara* 1981; *Cordo* and *Nashner* 1982; *Massion* 1984; *Hugon* et al. 1982; *Jung* 1982). Or, to cite *Menuhin* (1980): "It is a remarkable fact that the intentions of the violinist in his playing the violin is transposed not only in movements of his fingers but also in tiny movements of his back muscles." It is reasonable to assume that optimal performance of a delicate motor act, such as playing the violin, also requires optimization in postural support and correct adjustments. The commands for complex postural adjustments involving widely distributed muscles of the limbs and of the trunk must be issued at an early stage of movement initiation. Subsets of SMA neurones with their dominant control of trunk and shoulder muscles and with their relatively widely distributed efferent zones may well subserve such a role. Interestingly, it has been reported that patients with frontal lesions, but without paresis, may lack those anticipatory postural adjustments (*Elner* and *Gabibov* 1977). Firm experimental evidence for the participation of the SMA in this adjustment task is, however, lacking. It is also not inferred that the SMA is unique in this respect. Other structures have previously been considered to be involved, such as the premotor cortex (*Delacour* et al. 1972; *Freund* and *Hummelsheim* 1984) and the lateral cerebellum (*Massion* 1984). In monkeys, the caudal SMA is perhaps more involved in 'lower' controls (cf. Sect. 6.1), and the rostral SMA, more in 'higher' controls such as of vocalization (cf. Sect. 6.2).

7 Open Questions and Avenues for Future Research

In concluding this review, I should like to point to a number of problems which may be tackled in future studies on the SMA. The search for their solution will, I believe, provide further important insights into the functioning of the SMA.

7.1 Does the SMA Play a Crucial Role in Transmitting Learned Motor Programmes to the Motor Cortex?

Experience tells us that we perform our daily motor routines, or habits, in a rather stereotyped manner, i.e. according to a plan that has some invariant properties (*Viviani* and *Terzuolo* 1980). Intuitively, a motor learning process can be envisaged which gradually leads to a stable performance of learned motor acts. It is generally assumed that this stable motor performance is a result of *engram* formation; the stored programmes can then be called forth either by an 'internal' signal (drive) or by an appropriate external cue (trigger) signal.

Is there actual evidence of stored programmes? Which structure is the repository of a stored programme? How are stored programmes coded in the brain? How can these programmes be called forth and translated into a neural message? We are evidently still far from understanding these basic processes. That programmes are stored in the brain can hardly be denied, although the evidence for this is rather indirect. Thus, the pioneering brain-stimulation studies of *Hess* (1956) made it clear that strings of instinctive behaviour may be elicited artificially by electrical stimulation in discrete sites of the diencephalon. Stored experience can also be evoked by electrical stimulation within the hippocampus of man ('déjà-vu' experience). Lesions of various areas of the cerebral association cortex may lead to apraxia, a motor disorder which may be interpreted as a failure in motor programming. Neither the stimulation studies nor the lesion studies will, however, tell precisely in which structure the programme is located.

With respect to the SMA, previous lesion work in monkeys (cf. Chap. 4) indicates that the general drive to move in the environment and the plans for the normal movement repertoire are hardly affected by lesions. The described deficits are subtle, concerning perhaps the 'microstructure' of motor programmes. For example, it was noticed that slight changes occurred in the strategy with which objects were handled after unilateral lesions of the SMA (*Brinkman* 1984). Perhaps one of the most interesting observations was a peculiar loss of vocalization. A careful analysis of the typical patterns of vocalization of squirrel monkeys by *Kirzinger* and *Jürgens* (1982) revealed that a bilateral lesion of the SMA affected only the 'isolation peep', a call which is probably learned and not genetically determined. Perhaps the deficits in monkeys are mild because monkeys possess only a very limited repertoire of learned motor patterns that are of a symbolic nature.

Lesion work on the monkey's *premotor cortex* (lateral area 6) has produced a number of deficits, such as impairment of 'habits' (*Jacobson* 1934), impairment of postural adjustments (*Delacour* et al. 1972), impairment in visual reaching (*Moll* and *Kuypers* 1977), impairment of directed actions in response to different visual cue signals (*Halsband* and *Passingham* 1982), impairment of the ability to grasp food in the hemifield contralateral to the lesion, and hemi-inattention (*Rizzolatti* et al. 1983). It seems that the behavioural responses most affected by premotor lesions are those which depend on sensory (mainly visual) cues. Thus, one might conclude that the premotor cortex is important in retrieving learned motor programmes by means of *external* (mainly visual) cues.

It is now pertinent to inquire which specific programming function the supplementary motor area may subserve. The above-mentioned deficit in vocalization may be interpreted as an impairment of the ability to retrieve a learned motor program by *internal* signals or 'drives'. According to *Roland* et al. (1980b) and *Halsband* (1983), the SMA could also play a role in sequencing a string of learned motor acts. A change in the sequencing of picking up food morsels by SMA-lesioned animals was also seen by *Brinkman* (1984). Further studies in this direction are clearly needed. For this purpose, the monkey has to learn a relatively complex task and to execute it in a fairly stereotyped manner. Some technical difficulties may arise in sampling relevant parameters of movement trajectory, which make data reduction from recorded video tapes a necessity. The technique of reversible dysfunction by a cold-block (*Brooks* 1983) would be of particular help in such studies. A close analysis of the temporal structure of EMG patterns recorded from a number of different muscles should provide important clues to the role of the SMA in determining the temporal and spatial innervation patterns. It may even be worthwhile to investigate whether the well-known triphasic burst pattern (see e.g. *Sanes* and *Jennings* 1984) and the premovement suppression of muscle electromyographic activity (see e.g. *Conrad* et al. 1983), may change following an SMA lesion. Central to the hypothesis discussed in Sect. 6.3 is the question whether normal anticipatory postural changes occurring in many types of voluntary movements are lacking in SMA-lesioned animals.

7.2 Is the SMA of Particular Importance to Bimanual Coordination?

That the SMA may exert a bilateral control has been emphasized repeatedly, and this rests on anatomical data, single-unit recordings in a task requiring the right or the left arm, and lesion work with animals trained to perform a bimanual task (cf. *Brinkman* and *Porter* 1983). The remarkable deficit described by *Brinkman* (1981, 1984; cf. Sect. 4.5) should manifest itself when lesioned monkeys are investigated in their natural performance of handling objects with both hands, such as opening peanuts. *Pinto-Hamuy* (1956), who studied the retention of skilled movements after cortical ablations, ascribed the difficulty in handling objects to the occurrence of spasticity after lesioning of the SMA.

7.3 Does the SMA Contribute to the Rapid Initiation of Triggered Movements?

Despite the claim that the SMA may be specifically involved in the release of self-paced movements, it should be tested whether learned movements triggered by sensory cue signals are delayed after SMA lesions. It should be recalled that in the experiments of *Tanji* and co-workers (cf. Sect. 3.2) sensory trigger signals of various modalities were quite effective in driving SMA neurones. Surprisingly, conditioned movements in a reaction time situation have so far not been investigated in animals subjected to SMA lesions.

7.4 What is the Relative Timing of Neuronal Activation in the SMA and in the Precentral Motor Cortex?

Previous conflicting results on the timing of neuronal activation need to be clarified. In particular, the conditioned movement should be investigated in detail so that the picture of EMG patterns in distal, proximal, and trunk muscles becomes as complete as possible. Furthermore, the neurone populations of the SMA and of the precentral motor cortex should both be investigated with the same experimental paradigm and in the same monkeys. The problem bears on the important question whether the majority of neurones in the SMA are active 'upstream' from, or in parallel with, those of the motor cortex.

7.5 Concluding Remarks

As outlined in various sections of this review, investigations of the SMA of the human brain have shed new light on the function of the SMA. In contrast to observations made in SMA-lesioned animals, patients with lesions in the region of the SMA exhibit profound deficits in their overall drive to move in their environment, and spontaneous speech is markedly reduced. Cerebral blood flow measurements provide further support to the notion of a possible role of the SMA in the initiation of volitional movements, and recordings of premovement potentials also point in the same direction. Thus, these fascinating observations have led the discussion to the philosophical questions of will and intention (*Eccles* 1982; *Libet* et al. 1983a,b). Perhaps the most interesting results on the function of the SMA will come from clinical-physiological studies. Progress will presumably be slower in experimental research aimed at elucidating 'higher' functions of the SMA in animals.

The SMA of the human brain, together with other 'premotor' areas and the association cortex, has evolved to such a high degree that in many respects it is not comparable to the SMA of the monkey brain. However, the SMA of the monkey brain remains an extremely interesting cortical area, be it only for the fact that within that brain, it constitutes one of the major sources of afferent inputs to the motor cortex. In any attempt to understand the motor cortex one ought to take into account the discrete corticocortical links. One major issue of cortical neurobiology, in general, will be to understand the functional meaning of the corticocortical 'slabs'. They must be important organizing principles of the cortex. With respect to the motor cortex, the question arises how the somatotopical principle of organization is matched with the corticocortical interconnections with the SMA.

Acknowledgements. I thank my collaborators who participated in various stages of the research described in this review, in particular my wife Rita, who contributed so much with the anatomical investigations and with the illustrations. I thank also Mrs. P. Besson-Schouwey for her expert help, and Mrs. S. Rossier for efficient secretarial assistance. I am also grateful to Sir John Eccles for the many opportunities to discuss with him current work on the SMA. Dr. E. Rosenberg kindly helped in improving the English. My research described in this review was supported by the Swiss National Science Foundation (grants 3.752-80 and 3.522-83), the Swiss Multiple Sclerosis Society, and the Stanley-Thomas Johnson Foundation.

References

Adie WJ, Critchley M (1927) Forced grasping and groping. Brain 50:142–170
Alexander MP, Schmidt MA (1980) The aphasia syndrome of stroke in the left anterior cerebral artery territory. Arch Neurol 37:97–100
Asanuma C, Thach WT, Jones EG (1983a) Cytoarchitectonic delineation of the ventral lateral thalamic region in the monkey. Brain Res Rev 5:219–235
Asanuma C, Thach WT, Jones EG (1983b) Distribution of cerebellar terminations and their relation to afferent terminations in the ventral lateral thalamic region of the monkey. Brain Res Rev 5:237–265
Asanuma C, Thach WT, Jones EG (1983c) Anatomical evidence for segregated focal groupings of efferent cells and their terminal ramifications in the cerebello-thalamic pathway of the monkey. Brain Res Rev 5:267–297
Asanuma C, Thach WT, Jones EG (1983d) Brainstem and spinal projections of the deep cerebellar nuclei in the monkey, with observations on the brain stem projections of the dorsal column nuclei. Brain Res Rev 5:299–322
Bailey P (1933) In: von Bonin G (1944) Architecture of the precentral motor cortex and some adjacent areas, fig. 36, p 63 of the Precentral Motor Cortex (PC Bucy ed). The University of Illinois Press, Urbana
Bancaud J, Talairach J, Geier S, Bonis A, Trottier S, Manrique M (1976) Manifestations comportementales induites par la stimulation électrique du gyrus cingulaire antérieur chez l'homme. Rev Neurol (Paris) 132:705–724
Bancaud J, Chauvel P, Buser P (1985) Participation of SMA to speech. In: The cerebral events in voluntary movement; the supplementary motor and premotor areas. Proceedings of the Ringberg Symposium. Exp Brain Res 58:A14
Biber MP, Kneisley LW, Lavail JH (1978) Cortical neurons projecting to the cervical and lumbar enlargements of the spinal cord in young and adult Rhesus monkeys. Exp Neurol 59:492–508
Boschert J, Hnik RF, Deecke L (1983) Finger movement versus toe movement-related potentials: further evidence for supplementary motor area (SMA) participation prior to voluntary action. Exp Brain Res 52:73–80
Bouisset S, Zattara M (1981) A sequence of postural movements precedes voluntary movement. Neurosci Lett 22:263–270
Bowker RM, Murray EA, Coulter JD (1979) Intracortical and thalamic connections of the supplementary sensory and supplementary motor areas in the monkey. Abstr Soc Neurosci 5:704
Brinkman C (1973) Split-brain monkeys: cerebral control of contralateral and ipsilateral arm, hand and finger movements. MD thesis, Erasmus University of Rotterdam, Rotterdam, pp 112–163
Brinkman C (1981) Lesions in supplementary motor area interfere with a monkey's performance of a bimanual coordination task. Neurosci Lett 27:267–270
Brinkman C (1982) Supplementary motor area (SMA) and premotor area (PM) of the monkey's brain: distribution of degeneration in the spinal cord after unilateral lesions. Neurosci Lett (Suppl) 8:36
Brinkman C (1983) Effects of bilateral supplementary motor area lesions in the monkey. Neurosci Lett (Suppl) 15:23
Brinkman C (1984) Supplementary motor area of the monkey's cerebral cortex: short- and longterm deficits after unilateral ablation, and the effects of subsequent callosal section. J Neurosci 4:918–929
Brinkman C, Porter R (1976) Activities of cells in the supplementary motor cortex of the monkey during performance of a learned motor task. Proc Aust Physiol Pharmacol Soc 7:88
Brinkman C, Porter R (1979) Supplementary motor area in the monkey. Activitiy of neurons during performance of a learned motor task. J Neurophysiol 42:681–709

Brinkman C, Porter R (1983) Supplementary motor area and premotor area of monkey cerebral cortex: functional organization and activities of single neurons during performance of a learned movement. In: Desmedt JE (ed) Motor control mechanisms in health and disease. Raven, New York, pp 393–420
Brodmann K (1904/1905) Beiträge zur histologische Lokalisation der Grosshirnrinde. IIIte Mitteilung: Die Rindenfelder der niederen Affen. J Psychol Neurol 4:177–226
Brodmann K (1925) Vergleichende Lokalisationslehre der Grosshirnrinde in ihren Prinzipien dargestellt auf Grund des Zellenbaues, 2nd edn. Barth, Leipzig, p 334
Brooks VB (1983) Study of brain function by local reversible cooling. Rev Physiol Pharmacol 95:1–109
Brozoski TJ, Brown RM, Rosvold HE, Goldman PS (1979) Cognitive deficits caused by regional depletion of dopamine in prefrontal cortex of Rhesus monkey. Science 205:929–932
Campbell AW (1905) Histological studies on the localization of cerebral function. Cambridge University Press, Cambridge, p 360
Chauvel P (1976) Les stimulations de l'aire motrice supplémentaire chez l'homme. Remarques concernant son organisation fonctionnelle. Thesis, University of Rennes, Rennes
Cheema S, Rustioni A, Whitsel BL (1983) Corticospinal projections from pericentral and supplementary cortices in macaques as revealed by anterograde transport of horseradish peroxidase. Neurosci Lett (Suppl) 14:62
Cheney PD, Fetz EE (1985) Comparable patterns of muscle facilitation evoked by individual corticomotoneuronal (CM) cells and by single intracortical microstimuli in primates: evidence for functional groups of CM cells. J Neurophysiol 53:786–804
Cheney PD, Fetz EE, Palmer SS (1985) Patterns of facilitation and suppression of antagonist forelimb muscles from motor cortex sites in the awake monkey. J Neurophysiol 53:805–820
Chusid JG, de Guttierez-Mahoney CG, Marguls-Lavergne MP (1954) Speech disturbances in association with parasagittal frontal lesion. J Neurosurg 11:193–204
Conrad B, Benecke R, Goehmann M (1983) Premovement silent period in fast movement initiation. Exp Brain Res 51:310–313
Cordo PJ, Nashner LM (1982) Properties of postural adjustments associated with rapid arm movements. J Neurophysiol 47:287–302
Coxe WS, Landau WM (1965) Observations upon the effect of supplementary motor cortex ablation in the monkey. Brain 88:763–772
Damasio AR, Van Hoesen GW (1980) Structure and function of the supplementary motor area. Neurology 30:359
Deecke L, Kornhuber HH (1978) An electrical sign of participation of the mesial 'supplementary' motor cortex in human voluntary finger movement. Brain Res 159:473–476
Deecke L, Scheid P, Kornhuber H (1969) Distribution of readiness potentials, premotion positivity, and motor potential of the human cerebral cortex preceding voluntary finger movements. Exp Brain Res 7:158–168
Deecke L, Boschert J, Weinberg H, Brickett P (1983) Magnetic fields of the human brain (Bereitschaftsmagnetfeld) preceding voluntary foot and toe movements. Exp Brain Res 52:81–86
Delacour J, Libouban S, McNeil M (1972) Premotor cortex and instrumental behavior in monkeys. Physiol Behav 8:299–305
Denny-Brown D (1966) The cerebral control of movements. Liverpool University Press, Liverpool
Denny-Brown D (1967) The fundamental organization of motor behavior. In: Yahr MD, Purpura DP (eds) Neurophysiological basis of normal and abnormal motor activities, Raven, Hewlett, pp 415–444

De Vito J, Smith OA (1959) Projections from the mesial frontal cortex (supplementary motor area) to the cerebral hemispheres and brainstem of the Macaca mulatta. J Comp Neurol 111:261–278

Dhanarajan P, Rüegg DG, Wiesendanger M (1977) An anatomical investigation of the corticopontine projection in the primate (Saimiri sciureus). The projection from motor and somatosensory areas. Neuroscience 2:913–922

Eccles JC (1982) The initiation of voluntary movements by the supplementary motor area. Arch Psychiatr Nervenkr 231:423–441

Elner AM, Gabibov GA (1977) Possibility of recovery of the postural synergy after removal of the brain tumour. Agressologie 18:69–73

Fetz EE (1981) Neuronal activity associated with conditioned limb movements. In: Towe AL, Luschei ES (eds) Motor coordination, vol 5. Handbook of behavioral neurobiology. Plenum, New York, pp 493–526

Foit A, Larsen B, Hattori S, Skinhoj E, Lassen NA (1980) Cortical activation during somatosensory stimulation and voluntary movement in man: a regional cerebral blood flow study. Electroencephalogr Clin Neurophysiol 50:426–436

Fox PT, Fox JM, Raichle ME, Burde RM (1985) The role of cerebral cortex in the generation of voluntary saccades: a positron emission tomography study. J Neurophysiol 54:348–369

Freund HJ, Hummelsheim H (1984) Premotor cortex in man: evidence for innervation of proximal limb muscles. Exp Brain Res 53:479–482

Fulton JF (1949) Physiology of the nervous system, 3rd ed. Oxford University Press, New York, p 667

Gallouin F, Albe-Fessard D (1973) Stéréotaxie chez le macaque éveillé: étude des cortex somatiques, moteurs et moteur supplémentaire. J Physiol (Paris) 67:274A

Gelmers HJ (1983) Non-paralytic motor disturbances and speech disorders: the role of the supplementary motor area. J Neurol Neurosurg Psychiatry 46:1052–1054

Gemba H, Sasaki K (1984) Distribution of potentials preceding visually initiated and self-paced hand movements in various cortical areas of the monkey. Brain Res 306:207–214

Gilden L, Vaughan HG, Costa LD (1966) Summated human EEG potentials with voluntary movement. Electroencephalogr Clin Neurophysiol 20:433–438

Goldberg G, Mayer NH, Toglia JU (1981) Medial frontal cortex infarction and the alien hand sign. Arch Neurol 38:683–686

Goldman PS, Brown RM (1981) Regional changes of monoamines in cerebral cortex and subcortical structures of aging Rhesus monkey. Neuroscience 10:177–187

Green JR, Angevine JB, White JC, Edes AD, Smith RD (1980) Significance of the supplementary motor area in partial seizures and in cerebral localization. Neurosurgery 6:66–75

Grünewald G, Grünewald-Zuberbier E (1983) Cerebral potentials during voluntary ramp movements in aiming tasks: In: Gaillard AWG, Ritter W (eds) Tutorials in event-related potentials: endogenous components. North-Holland, Amsterdam, pp 311–327

Grünewald-Zuberbier E, Grünewald G (1978) Goal-directed movement potentials of human cerebral cortex. Exp Brain Res 33:135–138

Guidetti B (1957) Désordres de la parole associés à des lésions de la surface interhémisphérique frontale supérieure. Rev Neurol (Paris) 97:121–131

Gurfinkel VS, Kots JM, Paltsev FI, Feldman AG (1971) In: Gelfand IM, Gurfinkel VS, Fomin SSV, Tsetlin ML (eds) Models of the structural functional organization of certain biological systems. MIT Press, Cambridge, pp 382–395

Haider M, Groll-Knapp E, Ganglberger JA (1981) Event-related slow (DC) potentials in the human brain. Rev Physiol Biochem Pharmacol 88:125–197

Halsband U (1983) Higher disturbances of movement in Macaca fascicularis following discrete neocortical ablations. Neurosci Lett (Suppl) 14:154

Halsband U, Passingham R (1982) The role of premotor and parietal cortex in the direction of action. Brain Res 240:368–372

Hamada I (1981) Correlation of monkey pyramidal tract neuron activity to movement velocity in rapid wrist flexion movement. Brain Res 230:384–389
Hess WR (1943) Teleokinetische und ereismatische Kräftesysteme in der Biomotorik. Helv Physiol Acta 1:C62–C63 (English translation in: Akert K (1981) Biological order and brain organization. Selected works of WR Hess. Springer, Berlin Heidelberg New York, pp 265–268)
Hess WR (1956) Hypothalamus and thalamus. Documentary pictures. Thieme, Stuttgart
Hines M (1936) The anterior border of the monkey's (Macaca mulatta) motor cortex and the production of spasticity. Am J Physiol 116:76
Hugon M, Massion J, Wiesendanger M (1982) Anticipatory postural changes induced by active unloading and comparison with passive unloading in man. Pfluegers Arch Ges Physiol 393:292–296
Humphrey DR (1979) On the cortical control of visually directed reaching: contributions by nonprecentral motor areas. In: Talbot RE, Humphrey DR (eds) Posture and movement. Raven, New York, pp 51–112
Humphrey DR, Gold R, Reed DJ (1984) Sizes, laminar and topographic origins of cortical projections to the major divisions of the red nucleus in the monkey. J Comp Neurol 225:75–94
Ingvar DH (1977) Functional responses of the human brain studied by regional cerebral blood flow techniques. Acta Clin Belg 32:68–83
Ingvar DH, Philipson L (1977) Distribution of cerebral blood flow in the dominant hemisphere during motor ideation and motor performance. Ann Neurol 2:230–237
Ingvar DH, Schwartz MS (1974) Blood flow patterns induced in the dominant hemisphere by speech and reading. Brain 97:273–288
Jacobsen CF (1934) Influence of motor and premotor area lesions upon the retention of skilled movements in monkeys and chimpanzees. In: Localization of function in the cerebral cortex. An investigation of the most recent advances. Res Publ Assoc Res Nerv Ment Dis 13:225–247
Jonas S (1981) The supplementary motor region and speech emission. J Communication Disorders 14:349–373
Jones EG (1981) Functional subdivision and synaptic organization of the mammalian thalamus. Int Rev Physiol 25:173–245
Jones EG, Coulter JD, Burton H, Porter R (1977) Cells of origin and terminal distribution of corticostriatal fibers arising in the sensory-motor cortex of monkeys. J Comp Neurol 173:53–80
Jones EG, Coulter JD, Hendry SHC (1978) Intracortical connectivity of architectonic fields in the somatic sensory, motor and parietal cortex of monkeys. J Comp Neurol 181:291–348
Jürgens U (1984) The efferent and afferent connections of the supplementary motor area. Brain Res 300:63–81
Jung R (1982) Postural support of goal-directed movements: the preparation and guidance of voluntary action in man. Acta Biol Acad Sci Hung 33:201–213
Jung R, Hufschmidt A, Moschallski W (1982) Langsame Hirnpotentiale beim Schreiben: die Wechselwirkung von Schreibhand und Sprachdominanz bei Rechtshändern. Arch Psychiatr Nervenkr 232:305–324
Kalaska JF, Caminiti R, Georgopoulos AP (1983) Cortical mechanisms related to the direction of two-dimensional arm movements. Relations in area 5 and comparison with motor cortex. Exp Brain Res 51:247–260
Kalil K (1978) Neuroanatomical organization of the primate motor system: afferent and efferent connections of the ventral thalamic nuclei. In: Otto D (ed) Multidisciplinary perspectives in event related brain potential research. US Government Printing Office, Washington
Kievit H, Kuypers HGJM (1977) Organization of the thalamo-cortical connexions to the frontal lobe in the Rhesus monkey. Exp Brain Res 29:299–322

Kim R, Nakano K, Jayaraman A, Carpenter MB (1976) Projections of the globus pallidus and adjacent structures: an autoradiographic study in the monkey. J Comp Neurol 169:263–289

Kirzinger A, Jürgens U (1982) Cortical lesion effects and vocalization in the squirrel monkey. Brain Res 233:299–315

Környey E (1975) Aphasie trancorticale et écholalie: le problème de l'initiative de la parole. Rev Neurol 131:347–363

Kornhuber HH, Deecke L (1965) Hirnpotentialänderungen by Willkärbewegungen und passiven Bewegungen des Menschen: Bereitschaftspotential und reafferente Potentiale. Pfluegers Arch Ges Physiol 284:1–17

Krieg WJS (1963) Connections of the cerebral cortex. Brain Books, Evanston, p 472

Kubota K (1985) Prefrontal and premotor contributions to the voluntary movement in learned tasks. In: The cerebral events in voluntary movement: the supplementary motor and premotor ares. Proceedings of the Ringberg Symposium. Exp Brain Res 58:A8

Kubota K, Hamada I (1979) Preparatory activity of monkey pyramidal tract neurons related to quick movement onset during visual tracking performance. Brain Res 168:435–439

Kubota K, Iwamoto T, Suzuki H (1974) Visuokinetic activities of primate prefrontal neurons during delayed-response performance. J Neurophysiol 37:1197–11212

Künzle H (1978a) An autoradiographic analysis of the efferent connections from 'premotor' and adjacent prefrontal regions (areas 6 and 9) in Macaca fascicularis. Brain Behav Evol 15:185–234

Künzle H (1978b) Cortico-cortical efferents of primary motor and somatosensory regions of the cerebral cortex in Macaca fascicularis. Neuroscience 3:25–39

Kurata K, Tanji J (1985) Contrasting neuronal activity in supplementary and precentral motor cortex of monkeys: II. Responses to movement triggering vs. nontriggering sensory signals. J Neurophysiol 53:142–152

Kutas M, Donchin E (1980) Preparation to respond as manifested by movement-related brain potentials. Brain Res 202:95–115

Kuypers HGJM (1981) Anatomy of descending pathways. In: Brooks VB (ed) Handbook of physiology, sect 1, vol 2; motor control. Am Physiol Soc, Washington, pp 597–666

Lang W, Lang M, Kornhuber A, Deecke L, Kornhuber H (1983) Human cerebral potentials and visuomotor learning. Pfluegers Arch Ges Physiol 399:342–344

Laplane D, Orgogozo JM, Meininger V, Degos JD (1976) Paralysie faciale avec dissociation automatico-volontaire inverse par lésion frontale. Son origine corticale, ses relations aves l'AMS. Rev Neurol (Paris) 132:725–735

Laplane D, Talairach J, Meininger V, Bancaud J, Orgogozo JM (1977) Clinical consequences of corticectomies involving the supplementary motor area in man. J Neurol Sci 34:301–314

Lassen NA, Ingvar DH (1972) Radioisotopic assessment of regional cerebral blood flow. Prog Nucl Med 1:376–409

Lemon RN, Hanby JA, Porter R (1976) Relationship between the activity of precentral neurones during active and passive movements in conscious monkeys. Proc R Soc Lond (Biol) 194:341–373

Libet B, Alberts WW, Wright EW, Lewis M, Feinstein B (1975) Cortical representation of evoked potentials relative to conscious sensory responses, and of somatosensory qualities in man. In: Kornhuber HH (ed) The somatosensory system. Thieme, Stuttgart, pp 291–308

Libet B, Wright EW, Gleason CA (1982) Readiness potentials preceding unrestricted 'spontaneous' vs. pre-planned voluntary acts. Electroencephalogr Clin Neurophysiol 54:322–335

Libet B, Gleason CA, Wright EW, Pearl DK (1983a) Time of conscious intention to act in relation to onset of cerebral activity (readiness-potential). The unconscious initiation of a freely voluntary act. Brain 106:623–642

Libet B, Wright EW, Gleason CA (1983b) Preparation – or intention – to act, in relation to pre-event potentials recorded at the vertex. Electroencephalogr Clin Neurophysiol 56:367–372

Macpherson JM, Marangoz C, Miles TS, Wiesendanger M (1982a) Microstimulation of the supplementary motor area (SMA) in the awake monkey. Exp Brain Res 45:410–416

Macpherson J, Wiesendanger M, Marangoz C, Miles TS (1982b) Corticospinal neurones of the supplementary motor area of monkeys. A single unit study. Exp Brain Res 48:81–88

Mark RF, Sperry RW (1968) Bimanual coordination in monkeys. Exp Neurol 21:92–104

Masdeu JC, Schoene WC, Funkenstein H (1978) Aphasia following infarction of the left supplementary motor area. Neurology 28:1220–1223

Massion J (1984) Postural changes accompanying voluntary movements. Normal and pathological aspects. Human Neurobiol 2:261–267

Matsumura M, Kubota K (1979) Cortical projection to the hand-arm motor area from post arcuate area in macaque monkey: a histological study of the retrograde transport of horseradish peroxidase. Neurosci Lett 11:241–246

Matsunami K, Hamada I (1981) Characteristics of the ipsilateral movement-related neuron in the motor cortex of the monkey. Brain Res 204:29–42

Matsunami K, Hamada I (1983) Activities of single precentral neurones of the monkey during different tasks of forelimb movements. Jpn J Physiol 33:309–322

Menuhin Y (1980) In: Daniels R (ed) Conversations with Menuhin. St Martin's, New York

Moll L, Kuypers HGJM (1977) Premotor cortical ablations in monkeys: contralateral changes in visually guided reaching behavior. Science 198:317–319

Muakassa KF, Strick PL (1979) Frontal lobe inputs to primate motor cortex: evidence for four somatotopically organized 'premotor' areas. Brain Res 177:176–182

Murray E, Coulter JD (1981) Organization of corticospinal neurones in the monkey. J Comp Neurol 195:339–365

Olesen J (1971) Contralateral focal increase of cerebral blood flow in man during arm work. Brain 94:635–646

Olszewski J (1952) The thalamus of Macaca mulatta. An atlas for use with the sterotaxic instrument. Karger, Basel

Orgogozo JM, Larsen B (1979) Activation of the supplementary motor area during voluntary movements in man suggests it works as a supramotor area. Science 206: 847–850

Palmer C, Schmidt EM, McIntosh JS (1981) Corticospinal and corticorubral projection from the supplementary motor area in the monkey. Brain Res 209:305–314

Pandya DN, Van Hoesen GW, Mesulam MM (1981) Efferent connections of the cingulate gyrus in the Rhesus monkey. Exp Brain Res 42:319–330

Penfield W, Jasper H (1954) Epilepsy and the functional anatomy of the human brain. Little, Brown, Boston

Penfield W, Welch K (1951) The supplementary motor area of the cerebral cortex. Arch Neurol Psychiatry 66:289–317

Petrides M, Pandya DN (1984) Projections to the frontal cortex from the posterior parietal region in the Rhesus monkey. J Comp Neurol 228:105–116

Pinto-Hamuy T (1956) Retention and performance of "skilled movements" after cortical ablations in monkeys. Johns Hopkins Hosp Bull 98:417–444

Porter R, Lewis MM (1975) Relationship of neuronal discharges in the precentral gyrus of monkeys to the performance of arm movements. Brain Res 98:21–36

Rizzolatti G, Matelli M, Pavesi G (1983) Deficits in attention and movement following the removal of postarcuate (area 6) and prearcuate (area 8) cortex in macaque monkeys. Brain 106:655–673

Roland PE, Larsen B, Lassen NA, Skinhoj E (1980a) Supplementary motor area and other cortical areas in organization of voluntary movements in man. J Neurophysiol 43:118–136

Roland PE, Skinhoj E, Lassen NA, Larsen B (1980b) Differential cortical areas in man in organization of voluntary movements in extrapersonal space. J Neurophysiol 43: 137–150
Roland PE, Meyer E, Shibasaki T, Yamamoto YL, Thompson CJ (1982) Regional cerebral blood flow changes in cortex and basal ganglia during voluntary movements in normal human volunteers. J Neurophysiol 48:467–480
Rolls ET (1983) The initiation of movements. In: Massion J, Paillard J, Schultz W, Wiesendanger M (eds) Neural coding of motor performance. Exp Brain Res (Suppl) 7:77–113
Rubens AB (1975) Aphasia with infarction in the territory of the anterior cerebral artery. Cortex II:239–250
Russel WR, Young RR (1969) Missile wounds of the parasagittal Rolandic area. In: Locke S (ed) Modern neurology. Little, Brown, Boston, pp 289–302
Sanes JN, Jennings VA (1984) Centrally programmed patterns of muscle activity in voluntary motor behavior of humans. Exp Brain Res 54:23–32
Sanides F (1968) The architecture of the cortical taste nerve areas in squirrel monkey (Saimiri sciureus) and their relationship to insular, sensorimotor and prefrontal regions. Brain Res 8:97–124
Schell GR, Strick P (1984) The origin of thalamic inputs to the arcuate premotor and supplementary motor areas. J Neurosci 4:539–560
Schlag J, Schlag-Rey M (1985) Unit activity related to spontaneous saccades in frontal dorsomedial cortex of monkey. Exp Brain Res 58:208–211
Schlag-Rey M, Schlag J (1984) Visuomotor functions of central thalamus in monkey. I. Unit activity related to spontaneous eye movements. J Neurophysiol 51:1149–1174
Sessle BJ, Wiesendanger M (1982) Structural and functional definition of the motor cortex in the monkey (Macaca fascicularis). J Physiol (Lond) 323:245–265
Shibasaki H, Barrett G, Halliday E, Halliday AM (1980) Components of the movement-related cortical potential and their scalp topography. Electroencephalogr Clin Neurophysiol 49:213–226
Smith AM (1979) The activity of supplementary motor area neurons during a maintained precision grip. Brain Res 172:315–327
Smith AM, Hepp-Reymond MC, Wyss UR (1975) Relation of activity in precentral cortical neurons to force and rate of force change during isometric contractions of finger muscles. Exp Brain Res 23:315–332
Smith AM, Bourbonnais D, Blanchette G (1981) Interaction between forced grasping and a learned precision grip after ablation of the supplementary motor area. Brain Res 222:395–400
Sutton D, Trachy RE, Lindeman RC (1981) Monkey vocalization: effects of supplementary motor damage. Soc Neurosci 7:240 (abstract)
Sutton D, Trachy RE, Lindeman RC (1985) Discrimination phonation in macaques: effects of anterior mesial cortex damage. Exp Brain Res 59:410–413
Talairach J, Bancaud J (1966) The supplementary motor area in man. Int J Neurol 5:330–347
Tanji J, Evarts EV (1976) Anticipatory activity of motor cortex neurons in relation to direction of an intended movement. J Neurophysiol 39:1062–1068
Tanji J, Kurata K (1979) Neuronal activity in the cortical supplementary motor area related with distal and proximal forelimb movements. Neurosci Lett 12:201–206
Tanji J, Kurata K (1981) Contrasting neuronal activity in the ipsilateral and contralateral supplementary motor areas in relation to a movement of monkey's distal hindlimb. Brain Res 222:155–158
Tanji J, Kurata K (1982) Comparison of movement-related activity in two cortical motor areas of primates. J Neurophysiol 48:633–653
Tanji J, Kurata K (1983) Functional organization of the supplementary motor area. In: Desmedt JE (ed) Motor control mechanisms in health and disease. Raven, New York, pp 421–431

Tanji J, Kurata K (1985) Contrasting neuronal activity in supplementary and precentral motor cortex of monkeys. I. Responses to instructions determining motor responses to forthcoming signals of different modalities. J Neurophysiol 53:129–141
Tanji J, Taniguchi K (1978) Does the supplementary motor area play a part in modifying motor cortex reflexes? J Physiol (Paris) 74:317–319
Tanji J, Taniguchi K, Saga T (1980) Supplementary motor area: neuronal response to motor instructions. J Neurophysiol 43:60–68
Tasker RR, Gentili F, Sogabe K, Shanlin M, Hawrylyshyn P (1975) Decorticate spasticity: a re-examination using quantitative assessment in the primate. Can J Neurol Sci 2:303–313
Thorpe S, Rolls ET (1982) Activity of supplementary motor area neurones during visual discrimination and feeding. Neurosci Lett (Suppl) 10:481
Toyoshima K, Sakai H (1982) Exact cortical extent of the origin of the corticospinal tract (CST) and the quantitative contribution to the CST in different cytoarchitectonic areas. A study with horseradish peroxidase in the monkey. J Hirnforsch 23:257–269
Travis AM (1955) Neurological deficiencies following supplementary motor area lesions in Macaca mulatta. Brain 78:257–269
Travis AM, Woolsey CN (1956) Motor performance of monkeys after bilateral partial and total cerebral decortications. Am J Phys Med 35:273–310
Van Buren JM, Fedio P (1976) Functional representation on the medial aspect of the frontal lobes in man. J Neurosurg 44:275–289
Viviani P, Terzuolo C (1980) Space-time invariance in learned motor skills. In: Stelmach GE, Requin J (eds) Tutorials in motor behavior. North Holland, Amsterdam, pp 525–533
Vogt BA, Pandya DN (1978) Cortico-cortical connections of somatic sensory cortex (areas 3, 1 and 2) in the Rhesus monkey. J Comp Neurol 177:179–192
Vogt C, Vogt O (1919) Allgemeinere Ergebnisse unserer Hirnforschung. J Psychol Neurol (Leipzig) 25:279–439
von Bonin G (1944) Architecture of the precentral motor cortex and some adjacent areas. In: Bucy IC (ed) The precentral motor cortex. University of Illinois Press, Urbana, pp 7–82
von Bonin G, Bailey P (1947) The neocortex of Macaca mulatta. University of Illinois Press, Urbana (Illinois monographs in medical sciences, vol 5), p 163
von Economo C, Koskinas GN (1925) Die Cytoarchitektonik der Grosshirnrinde des erwachsenen Menschen. Springer, Berlin, 810 pp
Walker AE (1940) A cytoarchitectural study of the prefrontal area of the macaque monkey. J Comp Neurol 73:59–86
Wiesendanger M (1981a) Organization of secondary motor areas of cerebral cortex. In: Brooks (ed) Handbook of physiology, the nervous system II, motor control. Am Physiol Soc, Washington, pp 1121–1147
Wiesendanger M (1981b) The pyramidal tract, its structure and function. In: Towe AL, Luschei ES (eds) Handbook of behavioral neurobiology, vol 5. Plenum, New York
Wiesendanger M, Wiesendanger R (1984) The supplementary motor area in the light of recent investigations. Exp Brain Res Suppl 9:382–392
Wiesendanger M, Séguin JJ, Künzle H (1973) The supplementary motor area – a control system for posture? In: Stein RB, Pearson KC, Smith RS, Redford JB (eds) Control of posture and locomotion. Plenum, New York, pp 331–346
Wiesendanger M, Rüegg DG, Lucier GE (1975) Why transcortical reflexes? Can J Neurol Sci 2:295–301
Wiesendanger M, Hummelsheim H, Bianchetti M (1985a) Sensory input to the motor fields of the agranular frontal cortex: a comparison of the precentral, supplementary motor, and premotor cortex. Beh Brain Res (in press)

Wiesendanger M, Hummelsheim H, Macpherson J (1985b) Microelectrophysiology of the supplementary motor area. In: The cerebral events in voluntary movement: the supplementary motor and premotor areas. Proceedings of the Ringberg Symposium. Exp Brain Res 58:A2

Wiesendanger R, Wiesendanger M (1985a) The thalamic connections with medial area 6 (supplementary motor cortex) in the monkey (Macaca fascicularis). Exp Brain Res 59:91–104

Wiesendanger R, Wiesendanger M (1985b) Cerebello-cortical linkage in the monkey as revealed by transcellular labeling with the lectin wheat germ agglutinin conjugated to the marker horseradish peroxidase. Exp Brain Res 59:105–117

Wise SP, Tanji J (1981) Supplementary and precentral motor cortex: contrast in responsiveness to peripheral input in the hindlimb area of the unanaesthetized monkey. J Comp Neurol 195:433–442

Woolsey CN, Settlage PH, Meyer DR, Spencer W, Pinto-Hamuy TP, Travis AM (1952) Patterns of localization in precentral and "supplementary" motor areas and their relation to the concept of a premotor area. Res Publ Assoc Res Nerv Ment Dis 30: 238–264

Woolsey CN, Erickson TC, Gilson WE (1979) Localization in somatic sensory and motor areas of human cerebral cortex as determined by direct recording of evoked potentials and electrical stimulation. J Neurosurg 51:476–506

Rev. Physiol. Biochem. Pharmacol., Vol. 103

Molecular Aspects of Band 3 Protein-Mediated Anion Transport Across the Red Blood Cell Membrane

HERMANN PASSOW

Contents

Max-Planck-Institut für Biophysik, 6000 Frankfurt am Main, W. Germany

1 Introduction

The hydrophobic lipid bilayer of the red cell membrane is nearly impermeable for hydrophilic solutes. Nevertheless, the transport of CO_2 in the blood involves a rapid exchange of HCO_3^- against Cl^- across the erythrocyte membrane. This exchange is facilitated by an integral membrane protein with a molecular weight of approximately 97 kDa which constitutes about 25%–30% of the total membrane protein. It can be easily localized on sodium dodecyl sulfate (SDS) polyacrylamide gel electropherograms, where it is found in the third major band from the top. This explains its designation: the band 3 protein.

Besides the mediation of anion exchange, other possible functions of the protein have been considered. There are suggestions that it is also involved in the transport of water (*Brown* et al. 1975; *Sha'afi* and *Feinstein* 1977; *Solomon* et al. 1982, 1983; *Benz* et al. 1984; *Yoon* et al. 1984; *Chasan* et al. 1984), sugars (*Shelton* and *Langdon* 1983; *Acevedo* et al. 1981), and cations (*Solomon* et al. 1982, 1983), but so far no general agreement seems to exist about their validity.[1] Other functions could possibly be related to the capacity of the glycoprotein's outward-facing carbohydrate moiety to react with lectins (*Findlay* 1974; *Tanner* and *Anstee* 1976) and of its inward-facing N-terminal segment to combine with hemoglobin (e.g., *Shaklai* et al. 1977; *Salhany* and *Shaklai* 1979; *Salhany* et al. 1980) and intracellular enzymes (*Strapazon* and *Steck* 1976, 1977; *Yeltman* and *Harris* 1980). There seems to be little doubt that it plays an important role in anchoring the constituents of the cytoskeleton in the lipid bilayer (*Bennett* and *Stenbuck* 1979, 1980; *Hargreaves* et al. 1980). These observations and the uncertainties about possible additional functions explain why the prosaic term "band 3" is usually preferred to the term "anion transport protein." Nevertheless, with exclusive reference to the anion transport function, *Wieth* and *Bjerrum* (1983) proposed to designate the band 3 protein "capnophorin," meaning carrier of smoke (i.e., CO_2).

The participation of some protein in anion transport across the red blood cell membrane was first inferred from the inhibitory action of proteolytic enzymes like pronase (*Passow* 1971; *Knauf* and *Rothstein* 1971) and papain (*Schwoch* et al. 1974), and it was suggested that the inhibition of anion transport by a number of amino-reactive reagents like 1-fluoro-2,4-dinitro-benzene (N_2ph-F) (*Passow* 1969) and 5-methoxy nitrotropone (*Passow* and *Schnell* 1969) was due to the modification of amino groups of these membrane proteins. In 1964, *Maddy* synthesized a nonpenetrating amino-reactive reagent called SITS (see Fig. 14). This compound was the isothiocyanate derivative of a stilbene disulfonic acid. The discovery of the participation in anion transport of the band 3 protein was greatly aided by the use of this and other nonpenetrating stilbene disulfonates, whose capacity to inhibit anion transport was first recognized by *Knauf* and *Rothstein* (1971). It was shown that these compounds bind nearly

---

1 A consensus has developed with respect to sugar transport. It is accomplished by band 4.5 (*Baldwin* et al. 1979; *Kasahara* and *Hinkle* 1977; *Carter-Su* et al. 1982; *Shanahan* 1982; *Shanahan* and *D'Artel-Ellis* 1984; *Deziel* et al. 1984; *Mueckler* et al. 1985, Science 229:941–945)

exclusively to the band 3 protein (*Cabantchik* and *Rothstein* 1974a; *Passow* et al. 1975), that there exists a linear relationship between inhibition and binding (*Lepke* et al. 1976; *Ship* et al. 1977), and that binding is complete when all of the $1.0-1.2 \cdot 10^6$ band 3 molecules per cell are occupied (*Passow* et al. 1975; *Zaki* et al. 1975; *Lepke* et al. 1976; *Ship* et al. 1977; *Halestrap* 1976). It was finally shown that a half turnover of the anion transport system moves 10^6 anions per cell across the membrane, indicating that virtually all stilbene disulfonate binding sites and hence virtually all band 3 molecules equally participate in the transport process (*Jennings* 1982b).

Insertion of purified band 3 protein into liposomes facilitates anion exchange (*Cabantchik* et al. 1980; *Ross* and *McConnell* 1978; *Köhne* et al. 1981, 1983; *Lukacovic* et al. 1981; *Darmon* et al. 1983). Although the preparations of the authors cited do not demonstrate a reconstitution of all the details of the kinetics of anion transport observed in the red cell, there is no doubt that in most of the published reconstitution work, the inserted band 3 protein accomplishes anion transport by a mechanism that is essentially similar to that in the intact cell membrane. Nevertheless, one should be aware that the observation of an enhanced anion transport does not guarantee a physiological mode of operation of the transport protein. For example, both lysozyme (*Kaplan* 1972) and glycophorin (*van Hoogevest* et al. 1983) mediate anion transport in liposomes. The transport mediated by the latter can even be inhibited by the stilbene disulfonate DIDS (see Fig. 14), albeit at concentrations several orders of magnitude higher than required to inhibit band 3-mediated transport.

In recent work by *Scheuring* et al. (1984, 1985), most of the hitherto existing limitations of reconstitution work quoted above have apparently been overcome. After liposome formation by dialysis of a lipid-band 3 mixture and removal of excess detergent by density gradient centrifugation, they obtained vesicle populations in which all incorporated band 3 molecules were oriented right side out. The anion flux was inhibitable by H_2DIDS or flufenamate at low concentrations similar to those required to inhibit anion transport in red cells. The kinetics of sulfate transport also showed the characteristic details known from studies in intact red cells: saturation kinetics, a pH dependence with a maximum near pH 6.2, and an activation enthalpy of about 30 kcal/mol.

Using the Sendai virus technique, band 3-containing liposomes or intact red cells could be fused with cells other than red cells that do not possess the anion transport protein. The incorporation of band 3 and the appearance of stilbene disulfonate-sensitive anion transport in the receiving cells could be demonstrated (*Cabantchik* et al. 1980). The incorporated band 3 is slowly degraded by the cells (*Beigel* and *Loyter* 1983).

The membranes of all types of cells studied so far are permeable for anions. In many instances saturable components of anion transport have been discovered that can be inhibited by stilbene disulfonates (*Kleinman* et al. 1981; *Langridge-Smith* and *Field* 1981; *Smith* et al. 1981; *Kimelberg* 1981; *Bentley* and *McGahan* 1980, 1982; *Bittar* et al. 1980; *Brodsky* et al. 1979; *Ehrenspeck* 1982; *Löw* et al. 1984; *Brown* and *Simmons* 1981; *Simmons* 1981), and *Kay* et al. (1983) have demonstrated the occurrence in many nucleated cells (human fibroblasts, neutrophils, mononuclear leukocytes, epithelial cells of the lung and the mouth, liver cells, certain carcinoma and neuroblastoma cells) of membrane proteins that are capable of binding antibodies against human band 3 protein. Nevertheless, the relationship between these other anion transport systems and that in the red cells is far from clear.

In all cases that have been studied in sufficient detail, observations have been made that do not easily fit into the picture obtained in red cells. For example, in Ehrlich tumor ascites cells, sulfate transport may be inhibited by stilbene disulfonates up to 80% without concomitant inhibition of Cl^- transport (*Levinson* 1982). This is quite different from the situation in human red cells, where the transport of these anion species is affected equally (*Ku* et al. 1979). In vesicles made from contraluminal membranes of the proximal tubular cells of rat kidney, a Na^+/SO_4^{2-} cotransport which does not occur in red cells has been observed, although a H^+/SO_4^{2-} cotransport, similar to the cotransport in mammalian red cells, may also exist (*Löw* et al. 1984). In those tissues, where the band 3 protein-mediated anion transport cannot be expected to predominate to the same extent as in red cells, a discrimination of saturable anion flux components from the anion flux mediated by the $Na^+/K^+/Cl_2^-$ cotransport system described by *Geck* et al. (1980) still needs to be accomplished.

A new approach to the question at hand comes from molecular biology. *Cox* et al. (1985) have shown that in the avian kidney a specific subset of kidney cells exists (the columnar epethelial cells of the proximal convoluted tubules of cortical nephrons) in which a peptide is synthesized which has a molecular weight greater than that of the erythroid band 3; it is synthesized by a mRNA distinct from the mRNA involved in erythroid development but nevertheless apparently derived from the same single gene as erythroid band 3. It is as yet unknown whether or not the transport kinetics of the two peptides differ.

There exists a considerable body of literature on anion transport and the band 3 protein, which is rapidly expanding and which has been reviewed repeatedly (*Cabantchik* et al. 1978; *Jennings* 1984, 1985; *Knauf* 1979, 1982, 1985; *Macara* and *Cantley* 1983; *Passow* 1982; *Rothstein* 1982, 1984; *Wieth* and *Brahm* 1985). The present review does not attempt to give an encyclopedic overview of the existing information. Instead, after a brief summary of the properties of the band 3 protein and its state in the lipid bilayer that surrounds the red cell emphasis will be placed on the presentation of selected observations and ideas about the relationships between the kinetics of anion transport and the structure of the transport protein, knowledge of which can be expected to facilitate the evaluation of the original papers on this subject, including the many papers that could not be dealt with in this review. First, the flux components of anion transport will be considered and discussed in terms of penetration across a channel with variable barrier heights as proposed by *Läuger* and his associates. This encompasses the previously considered Ping-Pong kinetics for anion exchange. Pertinent equations will be presented and applied to the problem of recruitment of the transport molecules into outward-facing or inward-facing conformations. Secondly, after this rather formalistic description of transport kinetics, the more specific questions about the chemical nature and arrangement of the amino acid residues involved in anion binding and translocation will be discussed. Thirdly, and finally, a brief summary will be provided of currently existing hypotheses about the molecular details of band 3 protein-mediated anion transport.

The studies of the molecular mechanisms of anion transport have interesting physiological implications with respect to CO_2 transport in the blood. These aspects will not be dealt with in this volume. Instead the reader is referred to a brief pertinent review by *Wieth* and *Brahm* (1980) and to further publications that deal with specific aspects of bicarbonate transport (*Wieth* 1979; *Wieth* et al. 1982a).

2 The Band 3 Protein: Properties and State in the Lipid Bilayer

2.1 Isolation of the Band 3 Protein and Formation of Its Major Proteolytic Fragments

There exist many convenient methods for the preparation of large quantities of reasonably pure band 3 protein (for review, see *Tanner* 1979). Most of them involve the use of ionic or nonionic detergents (see *Steck* 1974, 1978; *Steck* and *Yu* 1973; *Yu* et al. 1973), although purification without detergents (after dissolution in acetic acid) has also been employed (*Schubert* and *Domning* 1978). Derivatives that are soluble in aqueous media without detergents can be obtained by acylation of the protein (*Herbst* and *Rudloff* 1982, 1984). The easy access to the purified protein has stimulated attempts at peptide mapping (for reviews, see *Rothstein* 1982; *Macara* and *Cantley* 1983; *Knauf* 1985; *Jennings* 1985) and amino acid sequence analysis of some of the fragments (*Mawby* and *Findlay* 1982; *Brock* et al. 1983; *Kaul* et al. 1983).

On SDS polyacrylamide gel electropherograms, band 3 protein extends over a broad molecular weight range of at least 10 kDa. This diffuse appearance of the band disappears after treatment with suitable glycosidases (*Mueller* et al. 1979; *Fukuda* et al. 1979; *Jennings* et al. 1984), in agreement with the suggestion that it is largely due to the heterogeneity of the carbohydrate moiety of the molecule (*Drickamer* 1978). Such heterogeneity has already been demonstrated in early studies of the band 3 protein which showed that about one-fourth of the protein binds concanavalin A while the rest binds *Ricinus* lectin (*Jenkins* and *Tanner* 1977). The carbohydrates are bound to a single asparagine residue in the C-terminal 35-kDa region of the protein (*Drickamer* 1978). They form a single oligosaccharide chain of about 30 sugar residues, whose sequence has been studied by *Tsuji* et al. (1980, 1981) and *Fukuda* et al. (1984).

By proteolytic digestion in situ (*Drickamer* 1976, 1977), it is possible to split the band 3 protein into a number of well-defined fragments (Fig. 1). For the purpose of this review, it is necessary to recall (see *Steck* 1978; *Rothstein* 1982; *Macara* and *Cantley* 1983) that external trypsin produces no cleavage while internal trypsin splits off a hydrophilic 42-kDa piece (*Lepke* and *Passow* 1976; *Grinstein* et al. 1978), which is released from the membrane (*Steck* 1978) while the remaining 55-kDa piece stays in the bilayer. When the red cells are treated with external chymotrypsin, two fragments of 35 kDa and 60 kDa are formed that both remain associated with the membrane. When the chymotrypsin-treated membranes are isolated and then exposed to internal trypsin, the 55-kDa fragment described above is found to be cleaved into two fragments of 35 kDa and 17 kDa that continue to be associated with the membrane while the 42-kDa piece is detached. The chymotryptic-tryptic 17-kDa fragment is reduced to 14.5 kDa when the cleavage at the inner membrane surface is brought about by exposure to chymotrypsin rather than trypsin (*Ramjeesingh* et al. 1980a,b). A further reduction of the 35-kDa fragment to 9 kDa is achieved when the chymotrypsin is applied at higher concentrations and low ionic strength (*DuPre* and *Rothstein* 1981).

External papain splits the protein at three locations, two of which reside in close proximity to the chymotrypsin cleavage site that delinates the 17-kDa and 35-kDa segments. The four fragments formed include one which comprises about 0.7 kDa of the C-terminal end of the tryptic-chymotryptic 17-kDa fragment and a similarly small piece (six amino acid residues) of the 35-kDa segment. Apparently, this fragment is either degraded by the enzyme or released into the medium. The second fragment includes 65 amino acid residues of the N-terminal end of the 35-kDa chymotryptic fragment (*Jennings* et al. 1984) and essentially coincides with the peptic fragment of *Brock* et al. (1983) discussed below. The third fragment extends up to the C-terminus of the band 3 protein. In contrast

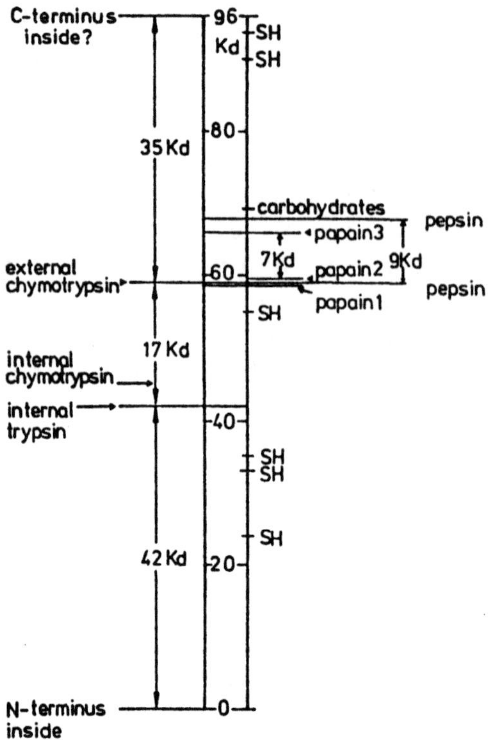

Fig. 1. Delineation of major proteolytic fragments as obtained after in situ digestion of human band 3 protein. For more detailed peptide maps, including smaller fragments, see *Ramjessingh* et al. (1983), *Jennings* (1985), and *Knauf* (1985). Location of the C-terminus is not yet unequivocally known, but possibly located at the inner membrane surface. Sequence information is available on: (1) 201 N-terminal amino acid residues of the 42-kDa fragment (*Kaul* et al. 1983). The fragment is hydrophilic, does not cross the bilayer and faces inward. (2) Several regions of the 17-kDa fragment, comprising about 36% of total amino acid content of that fragment (*Ramjeesingh* et al. 1980a; *Mawby* and *Findley* 1982), including N-terminal (*Jenkins* and *Tanner* 1977; *Markowitz* and *Marchesi* 1981; *Mawby* and *Findlay* 1982) and C-terminal (*Jennings* and *Adams* 1981) regions. The fragment is hydrophobic and crosses the bilayer at least three times (*Jennings* and *Nicknish* 1984). (3) The 9-kDa peptic fragment (72 amino acid residues, see Fig. 3) of the 35-kDa chymotryptic fragment. Alignment within the sequence is according to *Jennings* et al. (1984)

to former beliefs (*Jennings* and *Passow* 1979; *Jennings* and *Adams* 1981), the major fragments produced by papain remain hydrophobically associated with the membrane and are not released into the medium (*Jennings* et al. 1984).

Low concentrations of external pronase produce two fragments of 60 kDa and 35 kDa which, on polyacrylamide gel electropherograms, are indistinguishable from the fragments observed after treatment with chymotrypsin (*Cabantchik* and *Rothstein* 1974b). At higher concentrations or after longer times of exposure, the 60-kDa fragment is further digested and anion transport becomes gradually inhibited (*Passow* et al. 1977).

Mueller and *Morrison* (1977) discovered that among about 400 randomly selected blood donors there were 5%–6% mutants with red cells the treatment of which with pronase led to the appearance of a fragment of 63 kDa. Most of the individuals were heterozygotes, with band 3 molecules yielding equal amounts of 60-kDa and 63-kDa fragments; two homozygotes were found where only the latter fragments were produced. It is unknown whether or not there exist other mutants with differences of the sequence of the peptide chain that do not give rise to recognizable differences of molecular weight.

The hydrophilic 42-kDa segment carries the N-terminus, the hydrophobic 35-kDa segment the C-terminus of the band 3 protein. The former was found to be an acetylated methionine (*Drickamer* 1978). The absence of other end groups suggests the uniformity of the protein. This inference is supported by the establishment of peptide maps made from slices of the chymotryptic 35-kDa fragment that appears as a broad band of molecular weight 35 000–45 000 on SDS polyacrylamide gel electropherograms. Slices from different regions of certain subfragments yielded a unique tetrapeptide sequence (*Markowitz* and *Marchesi* 1981). These observations lend no support to speculations about the heterogeneity of the band 3 protein in homozygotes. The results further support the view (see p. 67) that the apparent heterogeneity of the band 3 protein or its 35-kDa C-terminal segment is entirely due to variability in glycosylation.

The proteolytic peptides described above could be highly purified (*Steck* 1978; *Steck* et al. 1978) and further cleaved by various chemical procedures, especially by bromocyanylation (for review, see *Rothstein* 1982). Peptides as small as 2 kDa have been found (*Ramjeesingh* et al. 1982). In this way, it was possible to obtain two fragments of the 14.5-kDa peptide mentioned above, one of 4.7 kDa, the other of 8.8 kDa. The N-terminal 38 amino acid residues of the latter piece have been sequenced (*Mawby* and *Findlay* 1982). Moreover, it was also possible to obtain by digestion with pepsin a fragment of the 35-kDa peptide of about 9 kDa whose sequence has been established (Fig. 3). A fragment of 1.8 kDa of this piece that is exposed to the inner cell surface carries a cluster of six positively charged groups, three lysines and three arginines (*Brock* et al. 1983). The 72 amino acid residues of the 9-kDa fragment sequenced by *Brock* et al. have a common N-terminus with the 35-kDa chymotryptic fragment (*Jennings* et al. 1984).

Peptide maps of the 42-kDa water-soluble fragment have been published (*Fukuda* et al. 1978) and the amino acid sequence of 201 N-terminal amino acid residues has been reported (*Kaul* et al. 1983).

```
1                                       11                                      21
MET GLY ASP MET ARG ASP HIS GLU GLU VAL LEU GLU ILE PRO ASP ARG ASP SER GLU GLU GLU LEU GLU ASN ILE ILE

                31                                      41                                      51
GLY GLN ILE ALA TYR ARG ASP LEU THR ILE PRO VAL THR GLU MET GLN ASP PRO GLU ALA LEU PRO THR GLU GLN THR

                                61                                      71
ALA THR ASP TYR VAL PRO SER SER THR SER THR PRO HIS PRO SER SER GLY GLN VAL TYR VAL GLU LEU GLN GLU LEU

        81                                      91                                      101
MET MET ASP GLN ARG ASN GLN GLU LEU GLN TRP VAL GLU ALA ALA HIS TRP ILE GLY LEU GLU GLU ASN LEU ARG GLU

                        111                                     121
ASP GLY VAL TRP GLY ARG PRO HIS LEU SER TYR LEU THR PHE TRP SER LEU LEU GLU LEU GLN LYS VAL PHE SER LYS

131                                     141                                     151
GLY THR PHE LEU LEU GLY LEU ALA GLU THR SER LEU ALA GLY VAL ALA ASN HIS LEU LEU ASP CYS PHE ILE TYR GLU

                161                                     171                                     181
ASP GLN ILE ARG PRO GLN ASP ARG GLU GLU LEU LEU ARG ALA LEU LEU LEU LYS ARG SER HIS ALA GLU ASP LEU GLY

                                191                                     201
ASN LEU GLU GLY VAL LYS PRO ALA VAL LEU THR ARG SER GLY GLY ALA SER GLU PRO LEU LEU PRO HIS GLN PRO SER

        211                                     221                                     231
LEU GLU THR GLN LEU TYR CYS GLY GLN ALA GLU GLY GLY SER GLU GLY PRO SER THR SER GLY THR LEU LYS ILE PRO

                        241                                     251
PRO ASP SER GLU THR THR LEU VAL LEU VAL GLY ARG ALA ASN PHE LEU GLU LYS PRO VAL LEU GLY PHE VAL ARG LEU

261                                     271                                     281
LYS GLU ALA VAL PRO LEU GLU ASP LEU VAL LEU PRO GLU PRO VAL GLY PHE LEU LEU VAL LEU LEU GLY PRO GLU ALA

                291                                     301                                     311
PRO HIS VAL ASP TYR THR GLN LEU GLY ARG ALA ALA ALA THR LEU MET THR GLU ARG VAL PHE ARG ILE THR ALA SER

                                321                                     331
MET ALA HIS ASN ARG GLU GLU LEU LEU ARG SER LEU GLU SER PHE LEU ASP CYS SER LEU VAL LEU PRO PRO THR ASP

        341                                     351                                     361
ALA PRO SER GLU LYS ALA LEU LEU ASN LEU VAL PRO VAL GLN LYS GLU LEU LEU ARG ARG ARG TYR LEU PRO SER PRO

ALA
```

Fig. 2a

Fig. 2. a Amino acid sequence of the 42-kDa domain of murine band 3 (residues 1–365) according to *Kopito* and *Lodish* (1985).
b Amino acid sequence of the 55-kDa domain of murine band 3 (residues 366–929). According to *Kopito* and *Lodish* (1985). The amino acid residues between the 2 horizontal lines are predominantly hydrophobic, those above and below the lines are predominantly hydrophilic. Accessibility to chemical modification (a) from the outside: Lys-449, Lys-558 (or Lys-561), Tyr-572, Tyr-646, (b) from the inside: Tyr-614, Cys-861, Cys-903, and (c) from the inside of the bilayer: Lys-608. The tryptic 55-kDa peptide (which forms the membrane-associated domain) comprises the 17-kDa peptide from Lys-366 to Tyr-572 and the 35-kDa peptide from Ala-573 to Val-929. The carbohydrates are presumably attached to Asn-660. The folding corresponds to the hydrophobicity plot in Fig. 2c and follows essentially the suggestions made by *Kopito* and *Lodish* except in the C-terminal region beyond Arg-800, where the folding was modified to take into account that Cys-861 and Cys-903 are believed to be located at or near the inner membrane surface. *Kopito* and *Lodish* assume that segment 11 penetrates all the way through the bilayer, that the hydrophobic region between Asp-825 and His-852 faces the outer surface and that the hydrophobic segment from Leu-853 to Leu-887 is again traversing the membrane. They leave it open whether the C-terminus faces the inside or the outside. The folding depicted in **b** serves to demonstrate the alternation of segments with high and low hydrophobicity. It does not necessarily represent the actual folding in the membrane and the horizontal lines do not necessarily coincide with the outer and inner surfaces of the lipid bilayer. The actual folding could, for example, include transmembrane segments with high hydrophobic momentum (as defined in *Eisenberg* 1984) rather than high hydrophobicity, as suggested in the figure

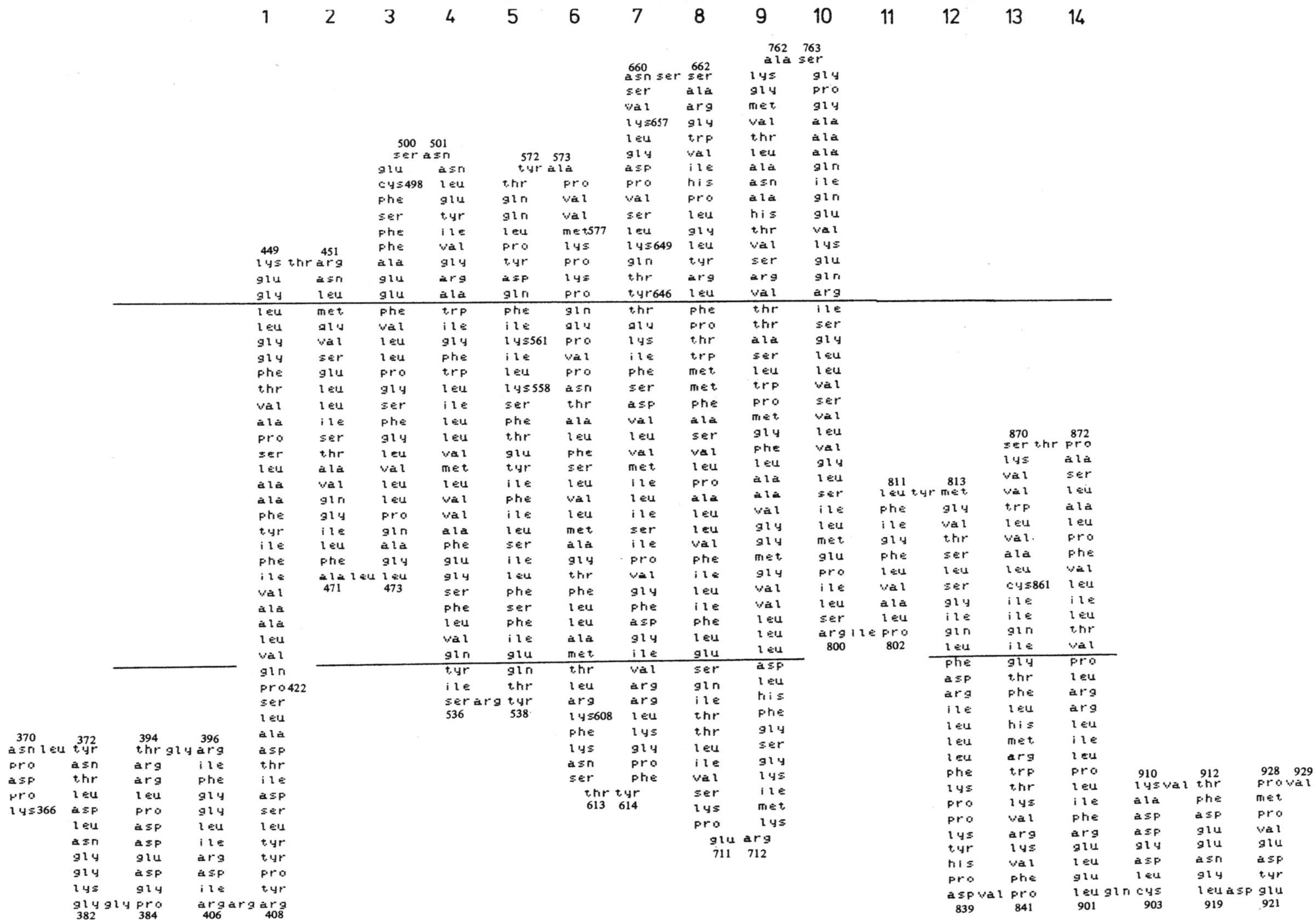

Fig. 2b

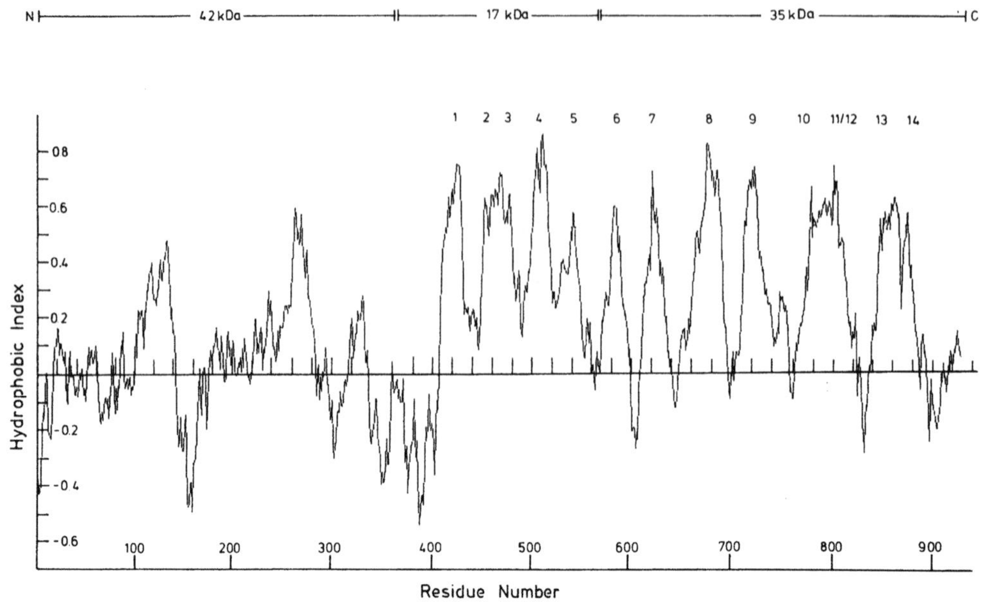

Fig. 2c. Hydrophobicity plot of the amino acid sequence of murine band 3. The hydrophobicities were averaged over windows of 21 amino acid residues. The hydrophobicity values assigned to the individual residues are the consensus values of *Eisenberg* (1984). Modified, according to *Kopito* and *Lodish* (1985). The numbers designating the various peaks refer to the transmembrane segments represented in Fig. 2b

It is noteworthy that in spite of the immense effort made in many competent laboratories, only minor pieces of the amino acid sequence of the hydrophobic transport-related domain of the band 3 protein have been established. One of the major reasons for the slow progress was due to the hydrophobicity of many of the cleavage products, which made it difficult to separate them from one another. Recently, however, *Kopito* and *Lodish* (1985) succeeded in isolating cDNA clones of the full length of the message for mouse band 3 protein. The sequencing of the cDNA does not encounter the difficulties involved in amino acid sequence analysis and was accomplished with standard procedures. The complete sequence obtained by these authors is represented in Fig. 2a,b.

There exists a considerable sequence homology between mouse and human band 3 (Fig. 3). The C- and N-terminal amino acid residues (valine and methionine, respectively) are the same in the two species, and the molecular weight of 103 000 daltons (corresponding to 929 amino acid residues) predicted from the sequence is close to the estimates of 90 000–100 000 obtained by SDS polyacrylamide gel electrophoresis of human and mouse band 3 protein. The sequence homology is least pronounced at the hydrophilic N-terminal 42-kDa portion of the peptide chain and quite striking for the pieces of amino acid sequence known to form the

```
                                10                                                    20
1  M  met gly asp met arg asp his — glu glu val leu glu ile pro asp arg asp ser glu glu glu leu glu
   H  met glu glu leu gln asp asp tyr glu asp —  — asp met —  —  —  —  —  glu glu asn leu glu

                    30                             40
      asn ile ile gly gln ile ala tyr arg asp leu thr ile pro val thr glu met gln asp pro glu ala leu
      —  —  —  —  gln glu glu tyr glu asp pro asp ile pro glu ser gln met glu glu pro ala ala his

        50                         60                                                     70
      pro thr glu gln thr ala thr asp tyr val pro ser ser thr ser thr pro his pro ser ser gly — gln
      asn thr glu ala thr ala thr asp tyr —  —  —  his thr — thr ser his pro —  —  gly thr his

                                            80                                   90
      — val tyr val glu leu gln glu leu met met asp gln — arg asn gln glu leu gln trp val glu ala
      lys val tyr val glu leu gln glu leu val met asp glu glu lys asn gln glu leu arg trp met glu ala

                              100                                  110
      ala his trp ile gly leu glu glu asn leu arg glu asp gly val trp gly arg pro his leu ser tyr leu
      ala arg trp val gln leu glu glu asn leu gly glu asn gly ala trp gly arg pro his leu ser his leu

                120                                   130                                   140
      thr phe trp ser leu leu glu leu gln lys val phe ser lys gly thr phe leu leu gly leu ala glu thr
      thr phe trp ser leu leu glu leu arg arg val phe thr lys gly thr val leu leu asp leu gln glu thr

                                         150                              160
      ser leu ala gly val ala asn his leu leu asp cys phe ile tyr glu asp gln ile arg pro gln asp arg
      ser leu ala gly val ala asn gln leu leu asp arg phe ile phe glu asp gln ile arg pro gln asp arg

                         170                                    180
      glu glu leu leu arg ala leu leu leu lys arg ser his ala glu asp leu gly asn leu glu gly val lys
      glu glu leu leu arg ala leu leu leu lys his ser his ala gly glu leu glu ala leu gly gly val lys

          190                                  200                                        210
      pro ala val leu thr arg ser gly gly ala ser glu pro leu leu pro his gln pro ser leu glu thr gln
      pro ala val leu thr arg ser gly asp pro ser gln pro leu leu pro gln his ser ser leu glu thr gln

      leu tyr cys
      leu phe cys

      375
2  M  leu asp leu asn gly gly
   H  leu asp leu asx gly gly

                         460                                   470
3  M  val ser glu leu leu ile ser thr ala val gln gly ile leu phe ala leu leu gly ala gln pro leu leu
   H  val ser glu leu leu ile ser thr ala val gln gly ile leu phe ala leu leu gly ala gln pro leu leu

      480                               490
      val leu gly phe ser gly pro leu leu val phe glu glu
      val val gly phe ser gly pro leu leu val phe glu glu

             580                                 590                                   600
4  M  pro lys pro gln gly pro val pro asn thr ala leu phe ser leu val leu met ala gly thr phe leu leu
   H  pro lys pro gln glu pro leu pro asn thr ala leu leu ser leu val leu met ala gly thr phe phe phe

                        610                                        620
      ala met thr leu arg lys phe lys asn ser thr tyr phe pro gly lys leu arg arg val ile gly asp phe
      ala met met leu arg lys phe lys asn ser ser tyr phe pro gly lys leu arg arg val ile gly asp phe

                  630                                    640
      gly val pro ile ser ile leu ile met val leu val asp ser phe
      gly val pro ile ser leu ile ile met val leu val asp —  phe
```

Fig. 3. Sequence homologies between mouse (*M*) and human (*H*) band 3 protein. According to *Kopito* and *Lodish* (1985). The partial amino acid sequences of human band 3 are from (*1*) *Kaul* et al. (1983) (*2*) and (*3*) *Mawby* and *Findlay* (1982) and (*4*) *Brock* et al. (1983). The numbers refer to the locations in the murine band 3 sequence

transmembrane portion of the peptide chain (see Fig. 3), including the C-terminal end where eleven of the twelve residues determined by carboxypeptidase Y digestion of human band 3 agree with the sequence predicted from the analysis of mouse cDNA.

The known amino acid sequence has been used in an attempt to explore the disposition of the band 3 protein according to the hydrophobicity of the various portions of the peptide chain. The "hydrophobicity plot" (*Eisenberg* et al. 1984) agrees with the notion that the hydrophilic 42-kDa segment (comprising in murine band 3 the 365 N-terminal amino acid residues) does not cross the lipid bilayer. The remaining portion of the peptide chain which is known to be associated with the lipid bilayer, shows 13 or 14 alternating regions of high and low hydrophobicity. The low hydrophobicity regions predominantly carry basic amino acid residues, except in the hydrophilic C-terminal region where there are no less than eleven aspartate or glutamate residues among the extreme 32 amino acid residues. *Kopito* and *Lodish* refer to this region as a third domain of the band 3 protein.

Although it is outside the scope of this volume, this section may be concluded by a brief summary of studies on the biosynthesis of band 3 protein. Publications from two independent research groups (*Lodish* and *Braell* 1982; *Braell* and *Lodish* 1981, 1982; *Sabban* et al. 1980, 1981; see also a review by *Sabatini* et al. 1982) report the successful expression of band 3 messenger RNA isolated from erythroid spleen cells of anemic mice in cell-free systems. When microsomal membranes are present, a cotranslational glycosylation and insertion into the lipid bilayer occurs. The primary translation product is not proteolytically processed, neither co- nor posttranslationally. The incorporation into the lipid bilayer seems to be initiated, therefore, by an internal signal sequence that, according to *Lodish* and *Braell* (1982), is located about 55%–60% of the distance from the N-terminus. This signal sequence can be expected to remain as a permanent transmembrane segment of the mature protein. Since the peptide chain crosses the membrane several times, multiple intrapeptidal signals for insertion, transfer, and stops may be involved. Once the band 3 protein is inserted into the endoplasmic reticulum (ER) membrane of the intact erythroid cell, it becomes transferred, virtually quantitatively, to the plasma membrane.

Messenger RNA isolated from erythroblasts of the spleens of anemic mice has been injected into prophase-arrested oocytes of *Xenopus laevis*. In five successful experiments, chloride transport was enhanced about threefold above the control level observed in uninjected oocytes. The additional flux could be inhibited by stilbene disulfonates (H_2DIDS, DNDS; see Fig. 14) while the control flux in the uninjected eggs could not be inhibited. In other experiments, the presence of band 3 protein in the cortices of the oocytes (which contain the plasma membrane) could be demonstrated by immunoprecipitation with antibodies against mouse band 3 protein. The binding of H_2DIDS to the inserted band 3 protein could not be demonstrated directly. Assuming that the turnover number per band 3 protein in the plasma membrane of the oocyte is about the same as in the red cell, one can estimate that the band 3 molecules sufficient for about 20 red cells became incorporated into the plasma membrane of each single oocyte. The specific activity of the $^3H_2$DIDS available would not suffice for the detection of this small number of band 3 molecules (*Morgan* et al. 1985).

In conclusion it may be added that posttranslational modification of band 3 in mature human red cells includes the methylation (presumably of D-aspartyl residues) by a cytosolic carboxyl methyltransferase [see *O'Connor* and *Clarke* (1983) for references] and the phosphorylation of a tyrosine residue in the 42-kDa cytoplasmic segment by a membrane-associated tyrosine kinase [see *Dekowski* et al. (1983) for references].

2.2 The Two Domains of the Band 3 Protein (Fig. 2a, 2b)

The hydrophobic 55-kDa fragment and the hydrophilic 42-kDa fragment that are obtained after cleavage by internal trypsin form two rather independent domains of the band 3 protein with different functions (*Guidotti* 1980; *Lysko* et al. 1981; *Snow* et al. 1981). The former is essentially responsible for anion transport, the latter for the binding of cytoskeletal proteins and, possibly, intracellular proteins.

Theoretically, one would expect that 28–35 amino acid residues (about 3–3.7 kDa) are required to form a peptide chain that traverses the lipid bilayer once.[2] Thus, in principle, the peptide chain of the 55-kDa domain could penetrate the bilayer about 15 times. There is evidence to the effect that it actually does traverse the bilayer at least five times (*Tanner* et al. 1980; *Ramjeesingh* et al. 1984). This evidence is essentially derived from investigations of the peptic in situ digestion of the band 3 protein. While the membrane-bound products of the digestion by other enzymes include relatively large fragments, pepsin only produces products of about 4 kDa. This corresponds roughly to the size of peptides that can cross the bilayer only once and suggests that pepsin solubilizes all exposed portions of the band 3 protein while only the 4-kDa membrane-spanning

2 According to *Eisenberg* et al. (1984), 21 hydrophobic amino acid residues in α-helical arrangement suffice to traverse the lipid bilayer once. Some additional residues are required to form the loops outside the bilayer, which connect consecutive transmembrane segments of the peptide chain

fragments survive. Five different peptic membrane-bound peptides have been identified (*Ramjeesingh* et al. 1984).[3] New evidence (*Jennings* and *Nicknish* 1984) and a careful reevaluation of the data hitherto available by *Jennings* (1985) and *Knauf* (1985) suggests that the peptide chain crosses the lipid bilayer at least seven times, the tryptic-chymotryptic 17-kDa fragment three times (*Jennings* and *Nicknish* 1984) and the 35-kDa fragment four to five times (*Jennings* et al. 1984).

The hydrophobicity plot (*Kopito* and *Lodish* 1985, see Fig. 2c) would suggest that the peptide chain of the membrane-associated region of the band 3 protein may traverse the lipid bilayer 12–13 times, and hence more than the 7 times predicted on the basis of the existing information on the peptide itself and less than the 15 times predicted as the maximal number of crossings that is feasible for a peptide chain of 509 residues. Studies on the sidedness of the chemical and enzymatic modification of the band 3 protein in situ are required for independent verification of the number of crossings derived from the hydrophobicity plot. The currently available data are still inadequate, but nevertheless, some correlations can be established and some of the crucial questions that need to be answered can be pinpointed.

The membrane-associated region (Fig. 2b) is usually counted from the cleavage site for intracellular trypsin, which forms the N-terminus of the 17-kDa fragment and, in the mouse band 3, is located at Lys-366. There follows a highly charged hydrophilic piece of the peptide chain, up to Ala-419, which is rich in arginine, tyrosine, and aspartate residues. This piece is quite evidently located outside the bilayer at the inner membrane surface. The next amino acid residues up to Lys-449 are almost exclusively hydrophobic and thus are likely to reside inside the bilayer. They may form an α-helical structure, although two prolines, one at the inner surface (422), the other inside the bilayer (438), may more or less interrupt the regularity of the helix. Lys-449 itself is located at the outer membrane surface, as was demonstrated by reductive methylation with ^{14}C-formaldehyde and borohydride (*Jennings* 1982a). Met-454 represents the exofacial beginning of the partial amino acid sequence obtained by *Mawby* and *Findlay* (1982). It extends up to Glu-492, whose location relative to the bilayer is unknown. On the basis of the hydrophobicity plot, *Kopito* and *Lodish* suggest that the segment of the peptide chain from Met-454 to Glu-492 forms an α-helix which traverses the membrane and returns to the outer surface, thereby forming a hydrophobic loop within the bilayer and possibly without exposure of the peptide chain to the medium at the inner membrane surface.

3 The 20-kDa piece of *Brock* et al. (1983) mentioned on p. 69 has been obtained by less exhaustive peptic digestion. It penetrates twice across the lipid bilayer (see Fig. 2b)

Cys-498 is difficult to label (*Rao* 1979; *Ramjeesingh* et al. 1982; *Jennings* et al. 1984). It has been suggested, therefore, that it is located inside the bilayer, and the hydrophobicity data would suggest that it resides near the outer surface where the hydrophilic segment emerges, which ranges from Glu-499 to Arg-509. An intracellular location of Cys-498, however, cannot be entirely excluded, since "stripping" of the ghost membrane with NaOH, (which leads to the release of membrane proteins attached to the inner membrane surface; *Steck* 1974) facilitates the reaction of this residue with NEM (*Jennings* et al. 1984).

One of the two lysine residues 558 or 561 (probably 558) is identical to Lys *a,* which is involved in covalent bond formation with one of the isothiocyanate groups of H_2DIDS (*Jennings* and *Passow* 1979; *Rudloff* et al. 1983). This residue is accessible only from the outer membrane surface but seems to be located just inside the bilayer (see the discussion of the structure of the stilbene disulfonate binding site on p. 173ff.).

The cleavage of the peptide by external chymotrypsin occurs at Tyr-572, which forms the C-terminus of the tryptic 17-kDa fragment. This residue is located inside the small fragment which is digested or released after exposure to external papain and which extends from near Leu-568 as its N-terminal end up to Gly-583 as its C-terminal end. Beginning at Ala-573 and overlapping with this peptide fragment is the peptic fragment of 72 amino acid residues sequenced by *Brock* et al. (1983). This fragment loops across the membrane up to Glu-648, which, in the human red cell is replaced by a glycine residue (see Fig. 3). In the human red cell, Lys-608 is still in a hydrophobic environment while Tyr-614 is susceptible to iodination by the lactperoxidase method from the inside and hence is exposed to the inner surface (*Brock* et al. 1983). The C-terminus at Gln-648 exceeds the papain fragment P_7 (*Jennings* et al. 1984), which begins at the N-terminal Gly-583 and ends at the C-terminal Thr-647. Twelve amino acid residues beyond Thr-647 towards the C-terminus of band 3 is located Asn-660, the single attachment point for the carbohydrates (*Kopito* and *Lodish* 1985), which projects into the external medium. The peptide chain is still accessible to iodination by means of external lactperoxidase at Tyr-674, suggesting that the chain dips into the bilayer byond Arg-675, where a series of hydrophobic amino acid residues follows. In other words, the peptide chain emerges near Lys-558 at the outer face of the bilayer, forms a hydrophilic bulge of about 26 amino acid residues, and disappears again in the bilayer near Pro-584. It then forms a loop which contacts the inner surface, returns to the outer surface and does not enter the bilayer again before Tyr-674.

Beyond this point of the peptide chain, no further independent information is available that could support the inferences drawn from the hydrophobicity plot. The only exceptions are the cysteine residues 861 and 903. In the human red cell membrane both residues are located at the inner membrane surface (*Rao* 1979; *Ramjeesingh* et al. 1982). If the localization of these residues is the same in the mouse red cell membrane, it will be necessary to postulate that the hydrophilic C-terminal peptide chain from Cys-903 up to Val-929 faces the cell interior. The intracellular location of Cys-861 is difficult to reconcile with the existence of transmembrane helices 11 and 12 which, according to the hydrophobicity plot, could form a loop across the membrane. Perhaps the amino acid residues between Arg-800 and Asp-825 form a hydrophobic pocket which does not penetrate all the way across the lipid bilayer. If so, the hydrophilic region between Asn-825 and Glu-858 would protrude into the internal medium, leaving Cys-861 near the inner surface and accessible to modification by NEM. The subsequent 28 amino acid residues (up to Leu-887) could again form a loop in the bilayer which does not penetrate to the opposite surface. In accordance with the hydrophobicity plot, the C-terminal domain would then begin with the arginine residues 888 and 889. It would be exposed to the inner membrane surface with Cys-903 susceptible to modification by NEM from the inside.

The folding of the peptide chain according to the hydrophobicity plot (assuming that the chain from Arg-800 up to the C-terminal Val-929 does not penetrate to the outer surface of the bilayer) assigns 15 basic and 12 acidic groups to the outer surface, and 33 basic and 31 acidic groups to the inner surface. Inside the bilayer the plot locates two basic amino acid residues (both of which are lysine residues in transmembrane segment 5) and five acidic amino acid groups (two of which are located in transmembrane segment 5). Thus the chosen arrangement of the peptide chain suggests that there are no strongly amphiphilic transmembrane segments that are likely to form aqueous pores lined with charged groups. Also, neither the serine nor the threonine residues can be arranged to provide an uncharged, hydrophilic lining. Accepting that the folding occurs according to the hydrophobicity plot requires one to stipulate that anion transport across the bilayer would take place through pores with hydrophobic walls formed as the consequence of incomplete packing of adjacent transmembrane segments of the peptide chain. The existence of such pores is known for the gramicidin channels, although the structure of the pores across the band 3 protein would certainly be different. Nevertheless, one is left with the uneasy feeling that pores formed by the amphiphilic segments of the peptide chain, which the hydrophobicity plot locates in the surfaces of the bilayer, could also be assumed to constitute transmembrane segments provided their hydrophilic amino acid residues face one another while the hydrophobic amino acid residues face the lipids of the bilayer or the hydrophobic face of adjacent helices. Thus, much further work is required until more realistic propositions can be made concerning the disposition of the peptide chain in the bilayer. The disposition suggested by *Kopito* and *Lodish* (as modified above), nevertheless, does not seem to be unrealistic, since a folding of the peptide chain which would lead to the establishment of pores formed by amphiphilic segments of the peptide chain would probably require the exposure of strongly hydrophobic segments of the peptide chain to the interface between the bilayer and the adjacent aqueous media, or the existence of further, incompletely penetrating hydrophobic loops, similar to the loops suggested for segments 2 and 3, 11 and 12, 13 and 14.

Both the 17-kDa and the 35-kDa fragments of the 55-kDa domain are involved in anion transport (*Jennings* and *Passow* 1979). Moreover, cleavage by papain of the extrafacial 35-kDa chymotryptic fragment at distances of 6 and 7 amino acid residues from the N-terminus (*Jennings* et al. 1984) leads to an inhibition of anion equilibrium exchange (*Passow* et al. 1977; *Jennings* and *Passow* 1979) and reduces the affinity for noncovalent stilbene disulfonate binding (*Jennings* and *Adams* 1981; *Lieberman* and *Reithmeier* 1983). Both fragments contain targets for covalently binding chemical modifiers of anion transport. Thus, there exists at least one lysine residue in the C-terminal region of the 17-kDa fragment (probably Lys-558) that is located near the outer surface and combines with inhibitors such as isothiocyanate derivatives of stilbene disulfonates (*Ramjeesingh* et al. 1981; *Williams* et al. 1979; *Jennings* and *Passow* 1979; *Lieberman* and *Reithmeier* 1983), isothiocyano phenylsulfonates (*Mawby* and *Findlay* 1982) and 1-fluoro-2,4-dinitrobenzene (N_2ph-F) (*Rudloff* et al. 1983). The 35-kDa fragment also contains binding sites for amino-reactive modifiers of anion transport like H_2DIDS (*Jennings* and *Passow* 1979), pyridoxal phosphate (*Nanri* et al. 1983; *Matsuyama* et al. 1983), formaldehyde (*Jennings* 1982a), phenylisothiocyanate (Lys-608, *Kempf* et al. 1981), and N_2ph-F (*Rudloff* et al. 1983). In addition to the lysines, at least one transport-related arginine residue resides on the 35-kDa segment (*Bjerrum* et al. 1983). Modification with agents that combine preferentially with tyrosines (e.g., tetranitromethane) or histidines (diethyl pyrocarbonate, diazosulfanilic acid) either produced no effect or when they produced an effect (as did the two histidine reagents) their effects could not or at least not solely be attributed to a modification of the expected target groups (*Raida, Lepke,* and *Passow,* unpublished work). Carbodiimides inhibit, presumably by modification of carboxyl groups (*Andersen* et al. 1983), but their site of action is unknown. Although both the 17-kDa and 35-kDa fragments contain SH groups, SH reagents do not seem to affect anion transport.

In the native protein the 42-kDa fragment forms an elongated piece that protrudes into the cytosol (*Appell* and *Low* 1981). It contains three SH groups (*Rao* and *Reithmeier* 1978; *Rao* 1979), each of which is capable of forming an S-S bridge with the 42-kDa fragment of an adjacent band 3 molecule to constitute a dimer. Once one S-S bridge has been established, the other two SH groups are protected against further oxidation and hence no more than one cross-link can be formed (*Reithmeier* and *Rao* 1979).

The 42-kDa fragment binds the cytoskeletal protein ankyrin (*Bennett* and *Stenbuck* 1979) or cytosolic proteins like hemoglobin (*Salhany* et al. 1980), aldolase (*Murthy* et al. 1981a,b), glyceraldehyde-3-phosphate dehydrogenase (*Tsai* et al. 1982), phosphofructokinase (*Karadsheh* and *Uyeda* 1977; *Higashi* et al. 1979) and anti band 3 antibodies (*Low* et al. 1984).

Binding of the cytosolic proteins seems to take place near the N-terminus, where, within a piece of 33 amino acid residues, not a single one is basic and eight are acidic (*Kaul* et al. 1983). The minimal sequence structure involved in protein binding was further narrowed down by measuring the competition between peptides generated from the N-terminal 23-kDa fragment of the 42-kDa domain and the 42-kDa domain of the intact band 3 protein for a number of the cytosolic proteins that are known to bind to the band 3 protein (hemoglobin, aldolase, glyceraldehyde-3-phosphate dehydrogenase). The most effective competitor was an acid cleavage peptide of 23 amino acid residues, beginning at the (blocked) N-terminus and including 14 negatively charged amino acid residues (*Tsai* et al. 1982; *Murthy* et al. 1984). Protein binding is easily measurable only at low ionic strength and low pH, suggesting the involvement of ionic bonds. It is modulated by the presence of coenzymes and phosphoric acid esters (*Kliman* and *Steck* 1980a,b). It is not clear to what extent this binding takes place under physiological conditions.

There exists some evidence for hemoglobin binding to band 3 in the intact red cell (*Eisinger* et al. 1982) but direct measurements of hemoglobin binding to inside out vesicles showed no binding at physiological pH and ionic strength, even at high hemoglobin concentrations (*Wiedemann* and *Elbaum* 1983). The binding of glycolytic enzymes has also not yet been shown in the intact cell, although there exists evidence for the accumulation of certain of these enzymes near the Na-K pump (*Mercer* and *Dunham* 1982). For the glyceraldehyde-3-phosphate dehydrogenase, there are conflicting reports. The data of *Maretzki* et al. (1974) and *Saleemuddin* et al. (1977) suggest that binding at physiological pH and ionic strength is negligible. These authors attribute the appearance of the glyceraldehyde-3-phosphate dehydrogenase on SDS polyacrylamide gel electropherograms of the isolated membrane to an artifactual binding of the enzyme at the low ionic strength usually used for the preparation of the membranes. *Kliman* and *Steck* (1980a,b) observed that the release of the enzyme after hemolysis by saponin is slower than the release of hemoglobin. An analysis of the time course of reequilibration of the enzyme after hemolysis led them to the conclusion that, in the intact cell, about two-thirds of the glyceraldehyde-3-phosphate dehydrogenase is bound to the band 3 protein. *Kelly* and *Winzor* (1984) reinterpreted the previously published data about the binding of glyceraldehyde-3-phosphate dehydrogenase and aldolase on the assumption that these enzymes bind to the band 3 protein in their tetrameric forms. They think that the previously published conflicting claims about the binding of glycolytic enzymes to the band 3 protein under physiological conditions will be resolved when more information becomes available on the thermodynamic properties in concentrated solutions, notably on the covolumes of all reactants involved.

Although the question of the binding of native intracellular proteins to the band 3 protein under physiological conditons is still controversial, it has been postulated that at least denatured hemoglobin is capable of firmly associating with the 42-kDa domain in the intact red cell. This association is supposed to occur in the intact red cell after naturally occurring or phenylhydrazine-induced hemoglobin denaturation and to lead to a clustering of the band 3 molecules. These clusters provide the recognition sites for antibodies directed against senescent red blood cells. The bound antibodies are then responsible for the removal of the aged red cells from the circulation (*Low* et al. 1985).

Kopito and *Lodish* (1985) have discussed in detail the available information concerning the binding of cytosolic proteins, anti-band 3 antibodies and ankyrin to specific regions of the primary sequence of the 42-kDa fragment of band 3.

They pointed out that in human band 3, the binding of cytosolic proteins involves primarily the eleven N-terminal amino acid residues. They quote unpublished work of *V. Patel* that mouse band 3, in contrast to human band 3, does not bind glyceraldehyde phosphate dehydrogenase, and relate this finding to the lack of homology of the N-terminal amino acid sequence in human and mouse band 3 (Fig. 3).

Since the anti-band 3 antibodies do not interfere with the binding of hemoglobin, *Kopito* and *Lodish* propose that the antigenic region is located further away from the N-terminus. They refer to the exceptionally low homology of the region between residues 125–138 in band 3 of mouse and man. Since there is no cross reactivity between antibodies against mouse and human band 3, they believe that the region between the amino acid residues mentioned is likely to be involved in antibody binding.

The ankyrin binding site has not yet been identified. It has been suggested, however, by *Low* et al. (1984) that the site is close to the IgG binding site since anti-band 3 antibodies compete with ankyrin for binding.

Kopito and *Lodish* suspect that the ankyrin binding site is highly conserved among different species. They tentatively locate, therefore, the ankyrin binding site to the highly conserved region just above the IgG binding site.

A cleavage site for trypsin resides at Arg-194, in a prolin reach region which seems to form a hinge between two about equally large segments of the 42-kDa fragment. A similar hinge seems to exist at Lys-366, the site of tryptic cleavage that leads to the formation of the hydrophilic 42-kDa domain and the hydrophobic 55-kDa domain.

The cytoplasmic N-terminal fragment is reported to be released after trypsination as a dimer with a helical content of about 37%. Calorimetric measurements and determinations of tryptophane fluorescence suggest that the stability of the isolated fragment changes with pH and is largely controlled by a single ionizable group of apparent pK of 7.8 at 37°C. The released dimer retains the capacity to combine with the cytoplasmic proteins mentioned above, indicating that its tertiary structure did not change much after detachment from the membrane-associated C-terminal fragment. However, the intersubunit S-S bridge can no longer be established (*Appell* and *Low* 1981).

The absence of interactions between the two domains of band 3 protein has been confirmed by systematic studies of the variations of the heat capacity of the isolated red cell membrane as a function of temperature at a range of conditions that affect the C-terminal and N-terminal domains differently (*Appell* and *Low* 1982). It could be shown that the two domains can be denatured largely independently of one another. For example, after exposure of the isolated membrane to dilute NaOH or acetic acid (two procedures used to remove from the red cell membrane the peripheral membrane proteins (*Steck* 1974), different results are obtained. After NaOH treatment, the C-terminal domain retains its original stability while the N-terminal domain is denatured. After treatment with acetic acid, the inverse is found. Moreover, after H_2DIDS binding, the C-terminal domain is stabilized while the N-terminal domain retains its original sensitivity to denaturation. It should be noted, however, that the techniques employed are not sensitive enough to preclude the possibility of weak interactions. Thus the thermal stability of the enzymatically released 42-kDa fragment is slightly increased as compared with its stability in situ (*Appell* and *Low* 1982). We had the impression that the release by trypsin of the 42-kDa fragment in resealed red cell ghosts leads to a slight inhibition of anion transport (*Lepke* and *Passow* 1976), and it has been shown that the binding of H_2DIDS to the outward-facing C-terminal domain of band 3 increases the interaction of the internal N-terminal domain with hemoglobin (*Salhany* et al. 1980) and with ankyrin and spectrin (*Hsu* and *Morrison* 1983).

2.3 State of the Band 3 Protein in the Red Cell Membranes

It has been believed for many years that the band 3 protein exists in the red cell membrane as a stable, noncovalent dimer (*Steck* 1978). This belief was largely based on the observation that extraction of the red cell membrane with Triton X-100 yielded such dimers (*Clarke* 1975; *Yu* and *Steck* 1975; *Steck* et al. 1976). Recently, however, *Schubert* and his associates have demonstrated that these dimers are artifactual and the result of traces of contamination in the commercially available Triton. After extraction with highly purified Triton X-100 (*Schubert* et al. 1983) or other nonionic detergents (*Pappert* and *Schubert* 1982, 1983), they demonstrated by analytical ultracentrifugation the existence of reversible equilibria between monomers, dimers, and tetramers. This confirmed earlier observations of *Dorst* and *Schubert* (1979) about the sedimentation behavior of detergent-free isolated band 3 protein and led to the suggestion that similar equilibria could also exist in the red cell membrane. However, in the membrane, these equilibria should be influenced by the different lipid environment, by the well-known association of the band 3 molecules with the cytoskeleton by binding to ankyrin (*Hargreaves* et al. 1980) and by interactions with hemoglobin and other intracellular proteins that may combine with the 42-kDa domain of the band 3 protein (*Cassoly* 1983; *Cassoly* and *Salhany* 1983).

The distance between adjacent molecules of band 3 protein in the intact red cell membrane could be determined by resonance energy transfer measurements in the intact red cell membrane and was found to be 28–52 Å, which is much less than the calculated mean distance between unassociated monomers in the red cell membrane of about 75 Å (*Macara* and *Cantley* 1981a). This result was not unexpected since treatment of red cell ghosts with Cu^{2+} and *o*-phenanthroline (*Steck* 1972, 1974; *Reithmeier* and *Rao* 1979), or of intact red cells with the easily penetrating diamide (*Haest* et al. 1977), leads to an intermolecular cross-linking via disulfide bonds. After cross-linking, about 80% of the band 3 molecules can be extracted from the membrane in the form of covalent dimers (*Steck* 1974). Using other cross-linkers, *Wang* and *Richards* (1974, 1975) observed dimers and tetramers.

One problem of cross-linking experiments consists of distinguishing between cross-linked products of naturally occurring complexes from those cross-linked during accidental encounters between independent molecules in a fluid membrane. It is necessary, therefore, to initiate and complete the cross-linking reaction before such encounters can take place. This could be accomplished by flash photolysis of photosensitive heterobifunctional reagents (*Kiem* and *Ji* 1977), and the predominant occurrence of dimers was confirmed.[4] In addition, however, very large aggregates were found that contained, besides band 3 protein, spectrin and other membrane proteins. It remained unclear whether these aggregates consisted of a mixture of homoaggregates of the individual protein species or of heteroaggregates. When lipophilic photosensitive agents are used, above pH 5.5 no cross-links can be formed. However, below this pH, dimers and tetramers of band 3 protein were observed (*Mikkelsen* and *Wallach* 1976). The occurrence of tetramers has been invoked by *Macara* and *Cantley* (1981a) to explain some of their resonance energy transfer data mentioned above.

A preferential formation of dimers also takes place after cross-linking with the nonpenetrating bis(sulfosuccinimidyl)suberate and 3,3′-dithiobis-(sulfosuccinimidyl)propionate. When the red cells are treated with chymotrypsin prior to or after the exposure to these cross-linking agents, cross-linked products were obtained that always contained dissimilar subunits. Thus, the cross-links are always formed between the 35-kDa and 60-kDa chymotryptic fragments, regardless of whether the fragments originated from the same or an adjacent band 3 molecule. This would suggest that the two subunits of a dimer are oriented such that the 35-kDa piece of one of the subunits is closely associated with the 60-kDa piece of the other (*Staros* and *Kakkad* 1983).

Electron microscopic studies revealed the existence in the red cell of intramembrane particles which consist of band 3 protein (*Weinstein* et al. 1978) and, possibly, glycophorin (*Pinto da Silva* 1972; *Cherry* and *Nigg* 1980). The number of these particles is about one-third of the number of band 3 molecules per cell (*Weinstein* et al. 1980), compatible with the assumption that a considerable fraction of the band 3 molecules exist in the form of tetramers, but that also smaller oligomers do occur. Thus the results of this work are qualitatively consistent with the view that the association equilibria of the band 3 protein observed in solutions of nonionic detergents, involving monomers, dimers, and tetramers, also exist in the intact red cell membrane, where they are probably modified by interactions of the band 3 protein's 42-kDa domain with the cytoskeleton and the intracellular proteins (for review, see *Haest* 1982).

The interactions between the band 3 protein and the cytoskeleton are most easily demonstrable by the study of the diffusion of band 3 protein in the lipid bilayer (for reviews, see *Edidin* 1981; *Peters* 1981).

---

4 According to *Dorst* and *Schubert* (1979), little if any cross-linking takes place during an encounter, while cross-linking becomes extensive after an encounter leads to the formation of a complex between the protomers. Cross-linking removes the complex from the association equilibrium and thus leads to a continuous increase with time of the cross-linked oligomers

In the intact red cell membrane, the band 3 protein molecules undergo rotational diffusion around the axis perpendicular to the surface (*Cherry* et al. 1976). Two populations of rotation rates have been observed. The more rapid rotation of 60% of the band 3 molecules has been attributed tentatively to unhindered rotation in the lipid bilayer, the slower rotation of the remaining 40% to a hindrance imposed by a linkage of the band 3 molecules to cytoskeletal proteins, notably ankyrin and band 4.1 (*Nigg* and *Cherry* 1980). The population of freely rotating band 3 protein species is not uniformly sized and the size distribution varies with temperature and with the cholesterol-phospholipid ratio of the membrane (*Mühlebach* and *Cherry* 1982).

Lateral diffusion of the band 3 protein is barely detectable in the intact red cell membrane (*Peters* et al. 1974; *Kapitza* and *Sackmann* 1980) but becomes easily measurable after detachment of the cytoskeleton by low ionic strength or high temperature (*Golan* and *Veatch* 1980, 1982) or after proteolytic digestion of the cytoskeleton (*Passow* et al. 1986). The control of lateral diffusion by the cytoskeletal proteins is also most impressively demonstrated by a comparison of diffusion in normal red cells with red cells from mutants that are lacking the essential matrix proteins (*Sheetz* et al. 1980). Thus, in spectrin-deficient spherocytotic mouse red cells, the diffusion coefficient for the band 3 protein is about one-sixth that of the surrounding lipids and thus of the order of magnitude predicted by the theory of *Saffman* and *Delbrück* (1975) for the two-dimensional bilayer continuum. In the normal mouse erythrocyte, the lateral diffusion of the lipids is the same as in the spherocytotic cells, but the rate of lateral diffusion of the band 3 protein is only 1/300 of that rate (*Koppel* et al. 1981).

3 Kinetics of Anion Transport

The band 3 protein spans the red cell membrane. Its C-terminal 55-kDa domain is anchored in the aqueous phase of the outer medium by the carbohydrates that are attached to it. The N-terminal end resides in the hydrophilic 42-kDa domain which sticks out into the medium at the inner membrane surface. Hence anion transport cannot be accomplished by rotational diffusion of the transport protein across the lipid bilayer. The various segments of the peptide chain that traverse the lipid bilayer are very likely, therefore, to form a channel for the anions. Nevertheless, in contrast to the penetration through aqueous channels, anion transport represents essentially an exchange process whose contribution to the electrical conductance of the membrane is negligibly small under physiological conditions. Only under special experimental conditions can a flow component be observed that contributes to the conductance and could represent simple diffusion of the anion as such. We may stipulate, therefore, from the outset that conformational changes of a channel formed by the transport protein dominate the transport process. This may be described, for example, in terms of a theory of *Läuger* et al. (1980), by fluctuations of adjacent energy barriers in the channel which give rise to transport kinetics intermediate between those of a carrier and a channel with fixed barrier structure (*Fröhlich* 1984a).

Although this concept should apply to any anion species that is accepted by the transport protein as a substrate, there exist certain differences between the kinetics of monovalent and divalent anion transport. For this reason, it seems useful to discuss first the basic concepts of anion transport as studied with monovalent anions and then to add a section on the specifics of divalent anion transport.

The information on the molecular aspects of transport kinetics reviewed below is derived from studies of the time course of anion equilibration in cell suspensions. The calculation of the rate constants and fluxes is, thereby, based on the assumption that a suspension can be represented by a two-compartment system where the compartments are separated by a membrane of uniform properties with an anion-binding capacity that is small compared with the amount of anions present in the intracellular and extracellular compartments. The calculation does not take into account that the rate of anion equilibration depends on the surface/volume ratios and the permeabilities of the membranes of the individual cells. The application of the two-compartment concept requires that these quantities are essentially similar in different cells of the population. The fact that anion equilibration at Donnan equilibrium follows a simple exponential suggests that the statistical distribution of the transport-controlling factors is indeed rather narrow. Nevertheless, it is gratifying that recently two different techniques have been described that permit the direct measurements of anion transport in the individual cells of a population. Both involve the use of a fluorescent anion species. One technique measures the decrease of the fluorescence inside the cells in the time interval between mixing with a medium devoid of the fluorescent anion and passing through a flow tube that permits the detection of the fluorescence in individual cells after a given length of time. From this data point, the individual rate constants can be calculated (*Muirhead* et al. 1984). The other technique makes use of the method of fluorescence microphotolysis. The fluorescent anions inside a single red cell ghost are bleached by a laser beam and the subsequent entry of unbleached anions from the medium into the bleached ghost is followed photometrically. This yields the full time course of anion equilibration in an individual ghost. Using NBD-taurine (*N*-[nitrobenzo-2-oxal-1,3-diazole]-2-aminoethane acid) as the fluorescent anion species, it has been shown that repeated bleaching of the same ghost yields the same rate constant with a reasonably high accuracy (SD ± 15%) and that the equilibration rates in different ghosts can be grouped around a slightly skewed mononodal distribution curve. The anion influx can be inhibited by H_2DIDS. The pH and temperature dependence as determined by averaging measurements in single cells follows the same pattern as observed in whole cell suspensions. These results confirm that the flux measurements in red cell suspensions containing large numbers of red cells fairly closely reflect the average behavior of the individual cells (*Peters* and *Passow* 1984).

3.1 Monovalent Anions

3.1.1 Anion Transport With and Without Contribution to the Electrical Conductance

Conceptually, anion transport can be measured in two different ways. First, one can determine anion exchange. This can be done either with radioisotopes when the anion distribution is at Donnan equilibrium, or by following the net movements that occur when one anion species at one membrane surface exchanges against another species at the other surface. The former is called "self exchange" or "homo exchange" at equilibrium or, briefly, "equilibrium exchange"; the latter is called "net exchange" or "hetero exchange." Second, one can determine the net electrolyte movements across the membrane. For reasons of electrical neutrality, they involve the penetration of equivalent amounts of cations and anions, which depends on the permeabilities of both the cation and the anion. Such electrolyte movements are usually induced by the addition of valinomycin, which renders the KCl-containing red cells selectively permeable for K^+. The ensuing movements of K^+ ions down their electrochemical gradient into a NaCl-containing medium induce an electrical field which serves as a driving force for the movements of the accompanying anions. Using suitable assumptions about the mode of penetration (e.g., that it can be described by the Goldman equation) it is possible to calculate permeabilities for cations and anions separately. The anion transport that contributes to the conductance is usually called "anion net flow," or "conductive flow." The latter term seems preferable since it does not give rise to confusion between "net flow" and "net exchange," as defined above.

When the two types of measurements are carried out at 37°–38°C, pH 7.2–7.4, at chloride concentrations of 150–160 mmol liter^{-1}, one obtains an equilibrium exchange flux for the Cl^- ions of 5×10^{-8} mol cm^{-2} s^{-1} (*Brahm* 1977). The permeability coefficient for the net Cl^- efflux (P_{Cl}) amounts to 2×10^{-8} cm s^{-1} (*Hunter* 1971) from which one can calculate for zero membrane potential (i.e., the condition of equilibrium exchange) a flux of 2.2×10^{-12} mol cm^{-2} s^{-1}. Thus only 1 out of 2.3×10^4 Cl^- ions that cross the membrane can be driven by an electrical field and hence contribute to the electrical conductance of the red cell membrane. This indicates that most of the anion movements across the red cell membrane represent an "electrically silent" exchange rather than ionic diffusion (*Harris* and *Pressman* 1967; *Scarpa* et al. 1970).

Attempts have been made to corroborate the inferences drawn from the flux measurements by electrical measurements. Using microelectrodes, it was possible to confirm that, in the presence of valinomycin, the membrane potential was essentially a diffusion potential for K^+ rather than Cl^-, although the rate of a valinomycin-mediated K^+/K^+exchange as measured isotopically is far lower then the rate of Cl^-/Cl^-exchange (*Hoffman* and *Lassen* 1970; *Lassen* 1972, 1977). Similar results were obtained with potential-indicating cyanine dyes (*Hoffman* and *Laris* 1974). Recent determinations of the potential generated by the electrogenicity of the Na-K pump in red cells by means of the cyanine dyes yielded a conductance of about $1-2 \times 10^{-5}$ ohm^{-1} cm^{-2}. which corresponds to a permeability of about $2-4 \times 10^{-8}$ cm s^{-1} and thus comes close to what one would expect on the basis of the estimates of P_{Cl} described above (*Hoffman* and *Laris* 1984). When the cells are treated with H_2DIDS or when the Cl^- is substituted by SO_4^{2-}, lower values are obtained (*Hoffman* et al. 1980).

3.1.2 Band 3-Mediated Conductance Versus Band 3-Independent Conductance

The discovery of the two flux components has led to the hypothesis that there exist two parallel and independent pathways of anion transport: a band 3 protein-mediated "electrically silent" exchange and an independent diffusive flow across a parallel "conductance pathway."

This inference has been examined by several different methods (*Tosteson* et al. 1973; *Hunter* 1977; *Knauf* et al. 1977). For the present purpose it may suffice to quote experiments with stilbene disulfonates and a related compound called APMB (see Fig. 14) that inhibit anion transport by combining selectively with the band 3 protein. H_2DIDS was found to produce a partial inhibition of the anion conductance which varied parallel to the blockage of the anion exchange (*Knauf* et al. 1977). Moreover, the reversibly binding H_2DIDS analogue DAS (Fig. 14) which inhibits anion exchange only when applied to the outer membrane surface showed the same sidedness of action when applied to produce the partial inhibition of net Cl^- movements that contribute to the conductance of the red cell membrane. Finally, the reversibly binding disulfonate APMB was found to inhibit both Cl^- exchange and net flux from either membrane surface (*Kaplan* et al. 1976).

The partial inhibition of the anion conductance by the disulfonates mentioned above clearly suggested an involvement of the band 3 protein. The question remained, however, whether that fraction of anion conductance that persists after maximal inhibition by these agents is due to anion movements through an independent conductance pathway or to incomplete blockage of the pathway provided by band 3 protein.

Attempts to resolve this question included studies of the temperature (*Fröhlich* et al. 1983) and pH dependence (*Knauf* et al. 1983a) of the various flux components. Most transparent are the results of the study of pH dependence of Cl^- transport. Increasing the pH produces three distinct effects: the equilibrium exchange increases and passes through a flat maximum around pH 7.2; the H_2DIDS-sensitive net flux remains virtually independent of pH; and the H_2DIDS-insensitive net flux decreases continuously. These results strongly suggest that the H_2DIDS-sensitive and -insensitive net fluxes are mediated by two distinct pathways (*Knauf* et al. 1983a,b).

One may conclude, therefore, that the anion transport across the red blood cell membrane can be dissected into three components, two of which seem to be mediated by band 3 protein and another that is probably not. Band 3 may mediate an electrically silent exchange and a diffusive flow that contributes to the conductance. The other conductance comes from a band 3 protein-independent conductance pathway.[5]

3.1.3 Band 3-Mediated Conductance: Slippage Versus Ionic Diffusion?

The simplest explanation that could account for both the band 3 protein-mediated exchange and conductance would be represented by a model first envisaged by *Patlack* and later dealt with in great detail by many other investigators (*Patlack* 1957; *Jacquez* 1964; *Lieb* and *Stein* 1972, and others) (Fig. 7, p. 96).

Within the framework of this model an anion (*a*) that leaves the red cell would bind to the inward-facing conformation (*r*) of the transport protein. The complex between protein and anion (*ar*) would switch into an outward-facing conformation (*as*). The anion would now be released and the unloaded, outward-facing transport protein (*s*) would be available for anion binding and anion transport in the opposite direction. Such a model would permit exchange which does not contribute to the conductance and net anion movements that do contribute. The latter would occur if not only the loaded form of the transport protein could undergo the conformational change between inward-facing (*ar*) and outward-facing (*as*), but if after the release of the anion at the outer surface the unloaded transport protein (*s*) could return to the inward-facing conformation (*r*) where it would be available to pick up another anion. Such conformational changes of the unloaded transport protein are called "slippage," and it was originally supposed that the band 3 protein-mediated anion movements that contribute to the conductance are essentially due to this process.

5 Evidence has also been obtained for the occurrence of cotransport systems for K^+, Na^+, $2Cl^-$ and K^+, Cl^- in the membrane of the red cells of humans and other animal species which moves chloride together with alkali ions. These systems also accept Br^- but they are inhibited by many of the anion species that are transported by the band 3 protein (e.g., I^-, NO_3^-); for review see *Knauf* (1985)

Table 1. Apparent dissociation constants of anion binding to transfer and Dalmark's modifier site (mmol liter^{-1} at 0°C)

	Transfer site	Modifier site	Conditions	
			pH	Temperature
Fluoride	88	337	7.4	23°C
Chloride	67	335	7.2	0°C
Bromide	32	160	7.2	0°C
Iodide	10	60	7.2	0°C
Thiocyanate	3		7.4	4°C
Bicarbonate	10	600	8.7	0°C
Sulfate	30–40	350–600	7.2–7.4	25°C
Phosphate	68	200	7.2–7.8	25°C

Data taken from reviews of *Wieth* and *Bjerrum* (1983) and *Knauf* (1979) except for those for thiocyanate, which come from *Dissing* et al. (1981). Derived from measuring anion equilibrium exchange

In a simple transport system of the type described above, at high anion concentrations the transport protein becomes saturated with its substrate. Unloaded transport molecules are no longer available to carry out net flow and slippage should disappear. This prediction has been tested experimentally (*Knauf* and *Law* 1980; *Kaplan* et al. 1980, 1982, 1983).

Figure 4 illustrates the well-known fact that anion exchange is a saturable process (with some self-inhibition at high substrate concentration, Table 1). In contrast, both the H_2 DIDS-sensitive and H_2 DIDS-insensitive conductive flow continue to increase at substrate concentrations where the exchange is saturated. At the upper end of this concentration range, the latter two flow components may show some tendency to saturate, but it is clear that the saturating levels would be far outside the range where Cl^- exchange reaches saturation. These results show that under these conditions slippage cannot account for the net anion movements that contribute to the conductance. They suggest that both the band 3 protein-independent and the band 3 protein-mediated conductive flows represent ionic diffusion across membrane-traversing channels. It remains to be established whether or not the band 3 protein-mediated conductive flow observed under conditions where the transport system is unsaturated (*Fröhlich* et al. 1983) represents a contribution of slippage or of diffusive flow. *Fröhlich* (1983, 1984b) measured the rate of the band 3 protein-mediated conductive efflux in the presence of large anion concentration differences between the inside and outside of the red cells. Under these conditions, nearly all of the transport molecules are outward facing and are unloaded (see next section). The rate of efflux should now be entirely controlled by the rate of transition of the unloaded transport molecule from outward- to inward-facing conformation, i.e., by slippage. This

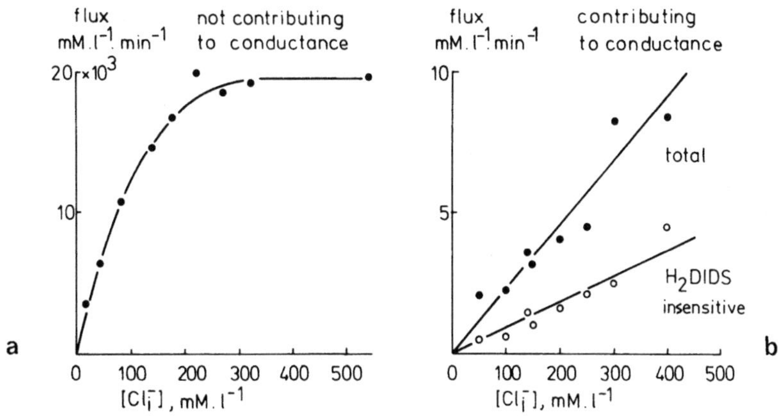

Fig. 4a,b. Concentration dependence of Cl^- equilibrium exchange **a** and conductive flow of Cl^- **b**. Equal Cl^- concentrations inside and outside human red cells. 30°C. **a** Date recalculated from ***Brahm*** (1977) for the temperature indicated. On the scale employed, inhibition by H_2DIDS would appear to be virtually complete. **b** Data calculated from the P_{Cl} values in Table 2 of ***Kaplan*** et al. (1983). ***Closed*** and ***open circles*** refer, respectively, to measurements in the absence or presence of 10 µmol liter^{-1} H_2DIDS: Note the 1000-fold difference between the ordinates

rate should be independent of the chemical nature of the transported anion species. In contrast to this prediction, Br^- and NO_3^- left the red cells 4.6 and 6.0 times faster, respectively, than Cl^-. This would suggest that even when band 3 is unsaturated, the contribution of slippage to anion conductance is small as compared with band 3 protein-mediated diffusion. Similar conclusions were drawn by *Knauf* et al. (1983a) from other types of studies.

3.1.4 Discussion of Band 3-Mediated Flux Components in Terms of a Channel with Variable Energy Barriers

At first glance it might seem somewhat surprising that even after saturation of its capacity to mediate anion exchange, the band 3 protein continues to mediate conductive anion flow. However, a superimposition of nonconductive and conductive flow has to be expected for ion movements across channels formed by protein molecules that are able to undergo conformational changes (*Läuger* et al. 1980; *Läuger* 1984).

Ion transport in channels has been described by a series of thermally activated jumps over a series of adjacent energy barriers with fixed heights. Models of this type were highly successful for the description of ion movements across channels in small peptides like gramicidin (see *Läuger* 1980, 1984). However, large-protein molecules at thermal equilibrium may exist in a large number of more or less easily interconvertible conformational states. It has been pointed out by *Läuger* and associates

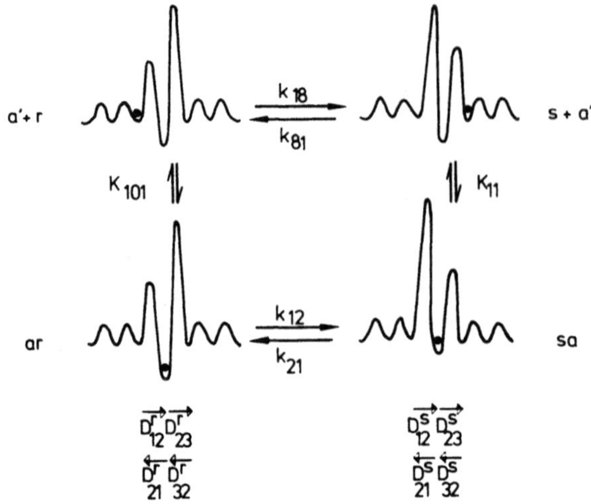

Fig. 5. Ion transfer across a channel with an alternating access gate characterized by two energy barriers of variable size. The transport kinetics are the resultant of (1) Ping-Pong kinetics for anion exchange as determined by the rate of conformational change of the gate and (2) the diffusive flow of the anion by jumps over the energy barriers without conformational change. The mass law equilibria for anion binding to the alternating access gate at the membrane surfaces $'$ and $''$ are designated by K_{101} and K_{11} respectively. The *k's* refer to the rate of transition of the energy barriers of the alternating access gate between facing the media at surfaces $'$ and $''$. In the calculations in the text it is assumed that the rate constants k_{12} and k_{21}, which pertain to the conformational transitions of the transport with a bound anion (*ar* $\rightleftharpoons$ *as*) and hence determine the rate of anion exchange, are large compared with the rate constants k_{18} and k_{81}, which describe "slippage", i.e., the conformational transitions of the transport protein without a bound anion. The rate constants for jumps of the anion over the energy barriers (diffusion) are denoted *D*. The indices *12,23* and *21,32* indicate the direction of the jumps; the indices *r* and *s* refer to the configurations of the energy barriers. Adapted from *Läuger* et al. (1980)

(*Läuger* 1980, 1984, 1985; *Läuger* et al. 1980) that for this reason the size of the energy barriers within the channel could be subject to thermal fluctuations. Both the rate of such fluctuations and the size of the barriers could be modified when an anion jumps into a potential well separating two barriers (i.e., a binding site). This may increase or reduce the height of the barrier in the diffusion pathway and thus inhibit or facilitate its own further movement (Figs. 5, 6). In the last chapter of this volume this will be considered in some detail in relation to specific models of anion transport which will be classified as "lock in" or "knock on" models, respectively.

The changes of the barrier heights that take place after the anion is bound and which describe the transition from inward-facing conformation (*ar*) to outward-facing conformation (*as*) are formally equivalent to carrier-mediated transport from the inner to the outer surface. The corresponding conformational changes of the unoccupied binding site would

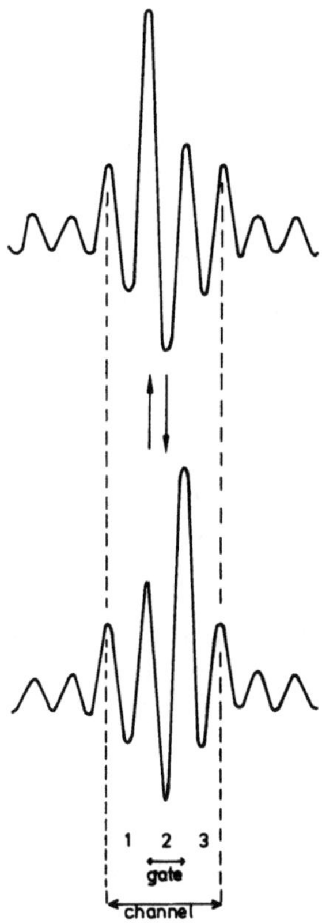

Fig. 6. Schematical representation of the extension of the model of anion transport across a channel with an alternating access gate by the incorporation of two approach sites, as suggested by *Tanford* (1985). The rate of anion exchange across the membrane depends on the probabilities of the occupancy of sites 1, 2, and 3. With increasing occupancy of site 2, the rate of exchange increases. With increasing occupancy of the approach sites 1 and 3, the probability for a release of an anion from site 2 decreases. Hence the rate of anion exchange passes through a maximum when the anion concentration in the adjacent media is increased (self-inhibition)

represent slippage and would be formally equivalent to the flow of the unloaded carrier. However, in contrast to the carrier-mediated transport, the channel with fluctuating barriers also permits jumps of a bound anion over the barriers (*Läuger* 1980) and thus encompasses the superimposed diffusional component, called "transit" by *Knauf* et al. (1983a) or "tunneling" by *Fröhlich* et al. (1983). The magnitude of this component would depend on (1) the ratio between inward- and outward-facing conformational states; (2) the barrier heights of each of the two conformational states, which in the most general case would give rise to different permeabilities for diffusive flow across inward-facing and outward-facing conformers; and (3) the relative probabilities for the jumping of the bound anion over the barriers, and the transition between the inward- and outward-facing barrier structures. The quantitative analysis of the model predicts that both the exchange due to the transitions $ar \rightleftharpoons as$ and the diffusion should saturate with increasing concentration of the substrate although not necessarily at the same concentration range.

A quantitative treatment of the available data in terms of this model has not yet been attempted but it seems clear, from the observations described above (see Fig. 5), that the model accommodates at least qualitatively the peculiar kinetics of anion exchange.

Since the diffusive flow across the channel is small compared with the electrically silent exchange mediated by the transitions $ar \rightleftharpoons as$, it is permissible for many purposes to neglect the diffusive component besides the exchange component. This will be done in Sect. 3.1.6, which will be devoted to the kinetics of the electrically silent net and equilibrium exchange.

3.1.5 Alternatives to the Model of a Channel With Variable-Energy Barriers

Although a channel with variable-energy barriers offers an attractive explanation for anion exchange and anion conductance from a unifying point of view, other interpretations of the findings cannot yet be ruled out. In particular it would seem feasible that the band 3 protein may oscillate between two different modes of operation, as a channel with more or less fixed barrier heights and as an exchanger. In the channel mode, the protein could give rise to single-channel events similar to other known channels, for example, the DIDS-inhibitable chloride channel from the electrical organ of Torpedo (see *Miller* 1984). These events would be associated with the diffusion of large numbers of anions, resulting in an easily measurable single-channel conductance. In the exchange mode, the protein could mediate an anion exchange that does not at all contribute to the electrical conductance.

Single-channel events have in fact been observed after incorporation of whole band 3 protein (*Benz* et al. 1984) or of the 17-kDa fragment of band 3 protein (*Galvez* et al. 1984) into planar lipid bilayers. The large conductances associated with the single-channel events of incorporated band 3 protein have been attributed to the formation of band 3 protein tetramers in the bilayer and are believed to be indicative of intermolecular channels that in the intact cell may be nonconducting and possibly responsible for water permeation (*Benz* et al. 1984). The origin of the single-channel conductance produced by the 17-kDa fragment is, apparently, still unclear. This fragment plays an important role in anion exchange, but in the intact cell membrane additional segments of the peptide chain are required to achieve proper functioning. Moreover, the reconstitution work quoted on p. 64 has amply demonstrated that proteins which in the intact cell membrane do not transport anions may do so when incorporated into lipid bilayers.

In our laboratory, the potassium permeability of the membranes, of intact red cells has been studied by means of the patch-clamp technique (*Grygorczyk* and *Schwarz* 1983; *Grygorczyk* et al. 1984). In the course of this work a large number of successful giga seals has been obtained, but anion channels of the size expected on the basis of the report of *Galvez* et al. (≈ 5 pS at 160 mmols liter^{-1}) could not be identified with certainty. Currently, attempts are being made to repeat the patch clamp experiments under special conditions, where the likelihood of the occurrence and discovery of anion channels is maximized (*Schwarz* and *Passow*, unpublished work). Such experiments have shown that anion-selective channels with a single-channel conductance of about 5 pS can indeed be observed. They are inhibitable by Persantine and possibly H_2DIDS, both of which are known to inhibit band 3-mediated net and exchange transport of anions. The number of the channels per cell is smaller by several orders of magnitude than the number of band 3 molecules and it is not yet clear whether or not they can account for the H_2DIDS-sensitive conductive anion flow which, as described above, has been attributed to band 3.

3.1.6 Anion Exchange

3.1.6.1 Modeling of Anion Transport: Equations for Equilibrium and Net Exchange in a Transport System Without Modifier Sites. Asymmetry Factors

The currently available informtion on that component of anion transport that does not contribute to the electrical conductance of the membrane is usually discussed in terms of a model which postulates anion binding to the transport protein and one or several subsequent conformational changes that result in the release of the anion at the other surface of the membrane.

At the present level of experimental resolution, a semiquantitative description of many observations can be obtained if the mathematical treatment of the model is based on three simplifying assumptions:

1. The transport rate is essentially governed by one single conformational change of the transport protein.

2. The conformational change leads to the transfer of one single substrate anion across the membrane. There occurs neither slippage (i.e., conformational change without anion binding and transport) nor an obligatory exchange of two or more anions bound to the same transport molecule at opposite membrane surfaces. The justification of the latter assumption has been discussed by *Knauf* (1979) and *Gunn* and *Fröhlich* (1982).

3. The rates of association and dissociation of the substrate are much higher than the rate of the conformational change that leads to the trans-

location of the bound anion. Hence, there exist equilibria at the two surfaces which can be described by the mass law.

A model based on these assumptions is depicted in Figs. 5 and 7. It gives rise to kinetics that had first been published by *Patlack* in 1957 before the discovery of the first transport protein. This model is essentially similar to the Ping-Pong model of enzyme kinetics in bulk solution described by *Cleland* (1963), where the enzyme substrate complex undergoes a conformational change before the products are formed and released. In contrast to an enzyme in bulk solution, however, the transport protein is located in a membrane that separates two media which may have different compositions. This gives rise to special phenomena that do not

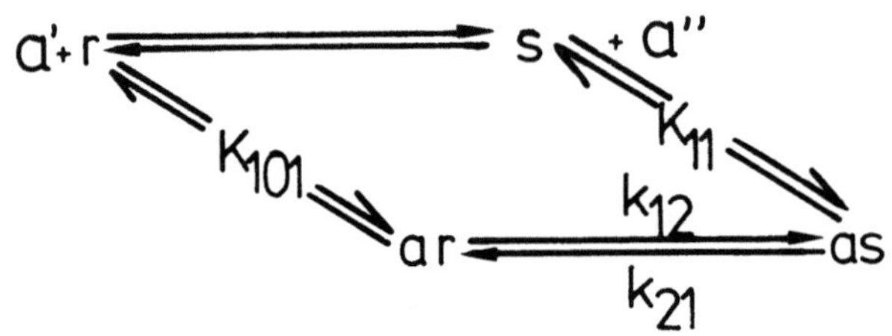

Fig. 7. Ping-Pong kinetics for one substrate, *a*. At surface ′, the substrate reacts with the *r* conformer of the transport protein to form the complex *ar*. The complex undergoes a conformational change into *as*, which enables the anion to be released at surface ″. The unoccupied conformer *s* is capable of combining with another anion *a* for the journey in the opposite direction. The formation of the substrate – protein complex is assumed to be fast compared with the rates of conformational transition $ar \rightleftharpoons as$, which are designated by k_{12} and k_{21}. For this reason, substrate binding is described by the mass law constants K_{101} and K_{11}. The transitions $r \rightleftharpoons s$ ("slippage") are conceivable and hence shown in the figure. The rates are assumed to be slow, however, compared with the rates of transition $ar \rightleftharpoons as$ and hence no rate constants are indicated

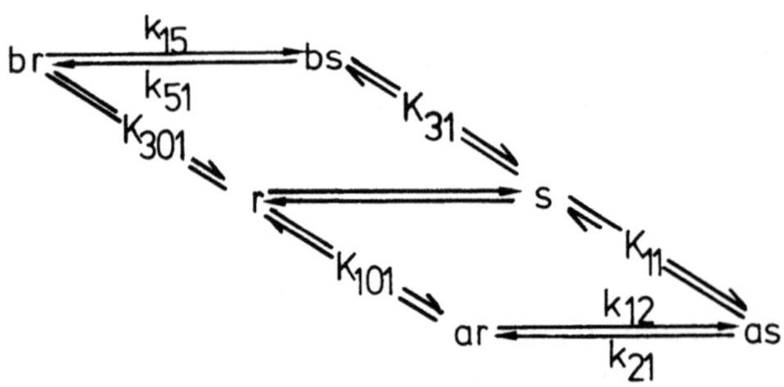

Fig. 8. Ping-Pong kinetics for two competing substrates, *a* and *b*. The transitions $r \rightleftharpoons s$ are not marked by rate constants to indicate that they are conceivable but assumed to be negligible compared with the transitions designated by the rate constants k_{12}, k_{21}, k_{15}, and k_{51}. K_{101}, K_{11}, K_{301}, and K_{31} designate mass law constants to describe binding to the transport protein in conformations *r* and *s*

occur in enzyme kinetics in bulk solution but which play an important role in the kinetics of membrane transport processes.

The present chapter serves two purposes: to provide a qualitative overview of the predictions of the model and to give a survey of the experimental work that has been interpreted on its basis. In addition, for readers who may find it convenient to have a more complete mathematical explication of the model at hand, derivations are supplied of the most widely applicable expressions along with their actual applications to specific experimental studies.

A mathematical treatment of the model depicted in Fig. 7 requires the derivation of expressions for:

1. The flux from one compartment to the other.
2. The concentrations of the various forms of the transport protein *ar, as, r, s,*

both as functions of the concentrations in the media at the two membrane surfaces (a' and a'') and of the parameter values defined in the reaction scheme.

Slippage will be neglected; the schema disregards, therefore, rate constants for the transitions $r \rightleftharpoons s$.

As will be shown below, the equilibrium exchange flux as defined by this model is described by the following equation:

$$j_{12} = \frac{k_{12}\,\overline{RS}}{1+q} \cdot \frac{a'a''}{a'a'' + \dfrac{K_{101}}{1+q}\,a'' + \dfrac{qK_{11}}{1+q}\,a'} \tag{1}$$

where $q = k_{12}/k_{21}$ and $\overline{RS}$ is the sum of all forms of the transport protein, i.e., $r + ar + s + as$.

The validity of this equation was demonstrated by flux measurements under three different experimental conditions: by varying a' at fixed a'', a'' at fixed a', and a at equal concentrations on both membrane surfaces: $a' = a'' = a$ (*Gunn* and *Fröhlich* 1979).

For all three sets of conditions, Eq. 1 can be transformed into expressions that are formally identical to the Michaelis-Menten equation. For each of the three sets of conditions one obtains, however, different values for the maximal transport rates and the half saturation constants, which will be designated respectively, V'_{max}, $K'_{1/2}$; V''_{max}, $K''_{1/2}$; and V_{max}, $K_{1/2}$. Since all of these constants are composite quantities (see Table 2), further information is required to derive from them k_{12}, k_{21}, K_{101}, K_{11}, i.e., the parameter values which define the system in terms of Ping-Pong kinetics.

Table 2. Ping-Pong kinetics of anion exchange at Donnan equilibrium

$$a' + r \underset{K_{101}}{\rightleftharpoons} ar \underset{k_{21}}{\overset{k_{12}}{\rightleftharpoons}} sa \underset{K_{11}}{\rightleftharpoons} s + a''$$

a varied, $a = a' = a''$	a' varied, a'' constant	a'' varied, a' constant
$j_{12} = V_{max} \dfrac{a}{a + K_{1/2}}$	$j_{12} = V'_{max} \dfrac{a'}{a' + K'_{1/2}}$	$j_{12} = V''_{max} \dfrac{a''}{a'' + K''_{1/2}}$
$V_{max} = \dfrac{k_{12}\,\overline{RS}}{1+q}$	$V'_{max} = \dfrac{k_{12}\,\overline{RS}}{1+q} \cdot \dfrac{a''}{a'' + \dfrac{qK_{11}}{1+q}}$	$V''_{max} = \dfrac{k_{12}\,\overline{RS}}{1+q} \cdot \dfrac{a'}{a' + \dfrac{K_{101}}{1+q}}$
$K_{1/2} = \dfrac{K_{101} + qK_{11}}{1+q}$	$K'_{1/2} = \dfrac{\dfrac{K_{101}}{1+q}\, a''}{a'' + \dfrac{qK_{11}}{1+q}}$	$K''_{1/2} = \dfrac{\dfrac{qK_{11}}{1+q}\, a'}{a' + \dfrac{K_{101}}{1+q}}$
$\dfrac{V_{max}}{K_{1/2}} = \dfrac{k_{12}\,\overline{RS}}{K_{101} + qK_{11}}$	$\dfrac{V'_{max}}{K'_{1/2}} = \dfrac{k_{12}\,\overline{RS}}{K_{101}}$	$\dfrac{V''_{max}}{K''_{1/2}} = \dfrac{k_{12}\,\overline{RS}}{qK_{11}}$

$$q = k_{12}/k_{21}$$

Much of what follows below is devoted to a description of the methods employed so far to obtain this additional information.

The explicit knowledge of the parameters can be used to predict the distribution of the various forms *ar, as, r,* and *s* between the membrane surfaces in the steady state. For a variety of conditions, asymmetrical distributions can be expected to occur. The ratios $(ar + r)/(as + s)$ and r/s are convenient measures of these asymmetries. They are numerical expressions for the so-called "recruitment" of the transport protein molecules into conformeric states where their transfer sites face compartments ′ (*r* and *ar*) and ″ (*s* and *as*) (see pp. 103, 106).

The reaction schema in Fig. 7 can be easily expanded to accommodate the interactions of two or more substrates that compete for the same transport system. For two substrates, the reaction schema assumes the form in Fig. 8.

The equations that describe the system show the competitive inhibition of one of the two anion species by the presence of the other (see Eqs. 4, 5, p. 102/103). They allow one to calculate flux and recruitment as functions of the various parameter values and of the concentrations a', a'', b', and b''.

It should be noted that under the conditions defined by the reaction schema in Fig. 8, net movements of both anion species *a* and *b* may occur. This "net exchange" or "hetero exchange" contrasts with the "homo exchange," which involves an exchange of anions of the same species. Hetero exchange may give rise to recruitment phenomena which are particularly pronounced when the penetration rates of the exchanging anion species differ considerably. They can be used to recruit the transport protein molecules into forms in which their "transfer sites" face either compartment $'$ or $''$ (see pp. 103, 106).

These transfer sites represent specific regions of transport protein molecules that are involved in substrate binding and translocation. They may be allosterically linked to many other regions of the transport molecule, the so-called modifier sites, which are capable of reacting with specific modifiers of transport. For example, in the band 3 protein the relationship between anion concentration and anion equilibrium exchange does not reach a plateau at high anion concentration as predicted by the Michaelis-Menten type kinetics derived from the model considered above. Instead, at high substrate concentrations self-inhibition is seen (Table 1). This has been attributed to substrate binding to an inhibitory modifier site (*Dalmark* 1975). Moreover, the transport rate varies with pH. Over the physiological pH range, these variations are related to the changes of the degree of dissociation of modifier sites that combine with protons. In addition there exist further modifier sites that react with noncompetitively acting inhibitors. The exact number of such sites is not yet known, but there are many of them.

The mathematical treatment of the influence of modifier sites can be accomplished by an extension of the reaction schemas in Figs. 7 and 8. For simplicity's sake, we shall first consider the case of two interacting substrate binding sites and a single substrate, *a*. The transfer site may be designated *r*, the modifier site *R*. At a given concentration a', the conformers with the transfer site facing compartment $'$ may exist in four different forms: *rR, arR, rRa, arRa* (the symbols represent respectively: combination neither with *r* nor *R;* with *r* or *R;* with both *r* and *R*). Each of these forms may undergo a conformational change that converts it into the corresponding conformer where the transfer site faces compartment $''$, viz., *sS, asS, sSa, asSa,* respectively. These conformers are at equilibrium with a''. Thus, instead of the four forms of the most simple model (*r, ar, s, as*), eight different forms need to be considered in this extended version.

For the graphical representation of the reaction schema of the extended model, it is useful to plot all forms with the transfer site facing compartment $'$ in one plane, and all corresponding forms with the transfer site

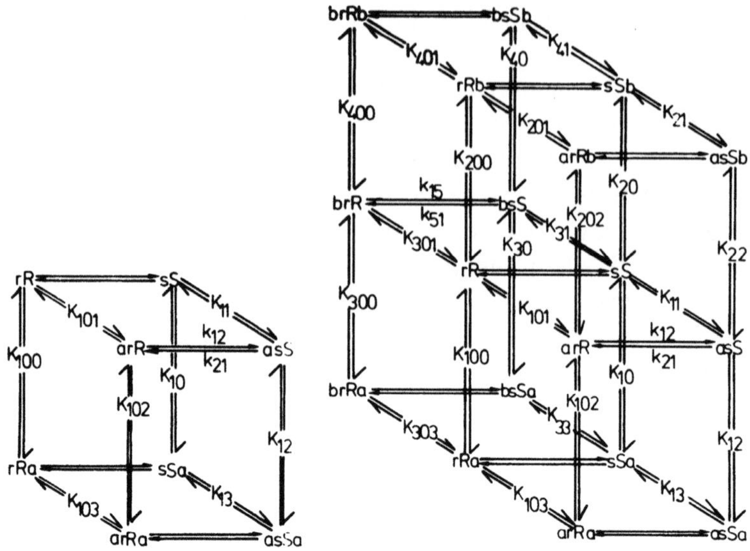

a b

Fig. 9. a Ping-Pong kinetics for one substrate, allosterically controlled by a second substrate binding site (a "modifier site") that exists in conformations R or S, depending on whether the transfer site faces compartment $'$ (r) or compartment $''$ (s), respectively. **b** Ping-Pong kinetics for two substrates, competing for the transfer site (r, s) and a modifier site (R, S). As in Figs. 7 and 8, the transitions only marked by *arrows* are conceivable, but assumed to be negligible compared with the transitions depicted by arrows and designated by the rate constants k_{12}, k_{21}, k_{15}, and k_{51}

facing compartment $''$ in another. This yields a three-dimensional structure which serves to define the various rate and equilibrium constants of the transport system (Fig. 9a).

When the transfer site alternates between the two membrane surfaces ($r \rightleftharpoons s$), the modifier site undergoes corresponding conformational changes ($R \rightleftharpoons S$), but without changing its orientation. Thus R and S are always exposed to the same solution and therefore at equilibrium with either a' or a''. In contrast, r is always exposed to a$'$, and s to a$''$. If it is assumed that the modifier site faces compartment $'$ containing the substrate at the concentration a', then the equilibrium exchange flux can be written:

$$j_{12} = \frac{k_{12}\,\overline{RS}}{1+q+\Gamma} \cdot \frac{a'a''}{a'a'' + \dfrac{K_{101}\,a'' + qK_{11}\,a'}{1+q+\Gamma}} \qquad (2)$$

where

$$\Gamma = \frac{K_{101}}{K_{100}} + \frac{a'}{K_{102}} + q\left(\frac{K_{11}a'}{K_{10}\,a''} + \frac{a'}{K_{12}}\right)$$

For $K_{10} = K_{100} = K_{12} = K_{102} \to \infty$ (i.e., no modifier site present) $\Gamma \to 0$ and Eq. 2 transforms into Eq. 1.

The three-dimensional schema in Fig. 9a can be readily extended to accommodate the actions of a second substrate b that competes with a for both the transfer site (in conformations r and s) and the modifier site (in conformations R and S). The reaction scheme appears as shown in Fig. 9b (*Passow* et al. 1980a). The equilibrium exchange flux is now described by the equation:

$$j_{12} = \frac{a'}{K_{101}} \cdot \frac{\overline{K}_S}{\overline{K}_S \overline{R} + q\overline{K}_R \overline{S}} \cdot k_{12}\overline{RS} \tag{3}$$

where

$$\overline{R} = 1 + \frac{a'}{K_{101}} + \frac{a'}{K_{100}}\left(1 + \frac{a'}{K_{103}} + \frac{b'}{K_{303}}\right) + \frac{b'}{K_{301}}\left(1 + \frac{b'}{K_{400}}\right) + \frac{b'}{K_{200}}\left(1 + \frac{a'}{k_{201}}\right)$$

$$\overline{S} = 1 + \frac{a''}{K_{11}} + \frac{a'}{K_{10}}\left(1 + \frac{a''}{K_{13}} + \frac{b''}{K_{33}}\right) + \frac{b''}{K_{31}}\left(1 + \frac{b'}{K_{40}}\right) + \frac{b'}{K_{20}}\left(1 + \frac{a''}{K_{21}}\right)$$

$$\overline{K}_R = \frac{a'}{K_{101}} + \frac{k_{15}}{k_{12}} \cdot \frac{b'}{K_{301}}\ ; \quad \overline{K}_S = \frac{a''}{K_{11}} + \frac{k_{51}}{k_{21}} \cdot \frac{b''}{K_{31}}$$

Figures 7, 9 serve to illustrate a method of representing diagrammatically transport systems of increasing complexity and to show that additional features can be incorporated relatively easily into the diagrams if:

1. The network of allosterical interrelationships between conformers with transfer sites facing toward one compartment are represented in one plane and those with transfer sites facing toward the other compartment in another plane.

2. Conformers with corresponding occupancies (e.g., *rR* and *sS; arR* and *asS*) are arranged in juxtaposition.

When these rules are followed, then all possible interrelationships will be considered automatically and the derivation of the equations for transport and recruitment can be accomplished with relative ease (see p. 102f. for further details).

The final equations, even those pertaining to the simplest possible model without modifier sites, are rather complex. It is a problem in its own right to find ways of writing them in forms that make their physical meaning as transparent as possible. It proved advantageous to follow two simple principles:

1. To avoid the condensation of individual constants into complex parameters whose physical significance is not obvious.

2. To use indices for the mass law constants K and the rate constants k which will allow one to recognize easily the symmetry of corresponding expressions for the two orientations of the transfer site. The mass law constants pertaining to anion or inhibitor binding to surfaces $'$ and $''$ are characterized by indices with three or two digit numbers, respectively, e.g., K_{101}, K_{11}; K_{301}, K_{31}, etc. The indices of the rate constants indicate the direction of the transitions. Pairs like k_{12}, k_{21} or k_{15}, k_{51} refer to rates of transitions from compartment $'$ to compartment $''$ and vice versa, respectively, whereby the first pair of this example refers to anion species a, the second to anion species b. Such notions help to recognize the significance of the individual parameters within the complex transport system and to identify mistakes in the derivations since the appearance of asymmetrical expressions usually indicates algebraic errors.[6]

Equations for Homo and Hetero Exchange in a Transport System Without Modifier Sites. Asymmetry Factors. Before discussing more complex situations, it is instructive to consider the predictions of Ping-Pong kinetics for a transport system without modifier sites.

These predictions can be derived from the following assumptions (the symbols are defined in Fig. 8).

Mass law equilibria at the two membrane surfaces:

$$a' \cdot r = K_{101} \cdot ar; \; a'' \cdot s = K_{11} \cdot as \qquad \text{(i)}$$
$$b' \cdot r = K_{301} \cdot br; \; b'' \cdot s = K_{31} \cdot bs$$

Steady state, no slippage:

$$k_{12}ar + k_{15}br = k_{21}as + k_{51}bs \text{ (see Appendix A)} \qquad \text{(ii)}$$

Mass conservation:

$$r + ar + br + s + as + bs = \overline{RS} \text{ where } \overline{RS} = \text{constant.} \qquad \text{(iii)}$$

The isotopically measured flux from compartment $'$ to compartment $''$ of anion species a (j_{12}) and b (j_{15}) is accomplished in a single step:

$$j_{12} = k_{12}ar \qquad \text{(4a)}$$
$$j_{15} = k_{15}br \qquad \text{(4b)}$$

6 After the establishment of the reaction network that represents all possible interactions between the sites that are believed to be involved in the given transport process, it is left to the experimentalist to elucidate which ones of the interactions predominate and which ones can be neglected. Thus the final reaction network that explains the experimental data may turn out to be less complex than the original network

Combination of Eqs. (i)–(iii) yields *ar* and *br* as functions of the composition of the media at the two membrane surfaces:

$$\frac{\overline{ar}}{\overline{RS}} = \frac{K''}{A'K'' + A''K'} \qquad \frac{\overline{br}}{\overline{RS}} = \frac{K_{101}\,b'}{K_{301}\,a'} \cdot \frac{\overline{ar}}{\overline{RS}} \tag{5}$$

where

$$A' = 1 + \frac{K_{101}}{a'} + \frac{b'K_{101}}{a'K_{301}} \qquad K' = k_{12} + k_{15} \cdot \frac{b'K_{101}}{a'K_{301}} \tag{5a}$$

$$A'' = 1 + \frac{K_{11}}{a''} + \frac{b''K_{11}}{a''K_{31}} \qquad K'' = k_{21} + k_{51} \cdot \frac{b''K_{11}}{a''K_{31}} \tag{5b}$$

Insertion of Eqs. 5, 5a, and 5b into Eqs. 4a and 4b, respectively, yields the fluxes of *a* and *b*. Eq. 5 is, however, also important in its own right since it can be extended to provide information on the molar fraction *p* of transport molecules whose substrate binding site faces compartment ′:

$$p = \frac{\overline{ar} + \overline{br} + \overline{r}}{\overline{RS}} = \left(1 + \frac{K_{101}}{a'} + \frac{K_{101}b'}{K_{301}a'}\right) \cdot \frac{\overline{ar}}{\overline{RS}}$$

(*Passow* et al. 1980b).

The remaining fraction of transport molecules $(1-p)$ faces compartment ″. Hence

$$p/(1-p) = (\overline{ar} + \overline{br} + \overline{r})/(\overline{as} + \overline{bs} + \overline{s}) = (A'K'')/(A''K') \tag{6a}$$

represents an asymmetry factor which does not distinguish between transport molecules with occupied and unoccupied substrate binding sites. Another asymmetry factor pertains to the transport molecules with unoccupied transfer sites.

$$\overline{r}/\overline{s} = (K''K_{101}a'')/(K'K_{11}a') \tag{6b}$$

(*Knauf* et al. 1980; *Knauf* 1982).

The numerical estimate of the parameter values that determine the flux $K_{101}, K_{11}, K_{301}, K_{31}, k_{12}, k_{21}, k_{15}, k_{51}$ (see Eqs. 4a, 4b) and calculation of the asymmetry factors (Eqs. 6a, 6b) constitute the final goal of the analysis of transport kinetics in terms of the Ping-Pong model of anion exchange.

As has already been pointed out (p. 97), for the numerical calculation of the parameter values from experimental data, it is usually convenient to transform Eqs. 4 and 5 into expressions that are formally identical to the Michaelis-Menten equation. This provides relationships between V_{max} and $K_{1/2}$ values defined by this equation and the various mass law and rate constants defined by the model. For a quantitative determination of these latter constants, it is necessary to take into account that, in general, they are interdependent and hence that not all of them can be fitted independently. For a discussion of this problem, see Appendix B.

Below some specific applications will be considered.

3.1.6.2 Equilibrium Exchange

Evaluation of the Mass Law and Rate Constants Defined by the Ping-Pong Model. The equation for homo exchange of a single anion species as discussed above (p. 97, Eq. 1) is obtained when $b' = b'' = 0$ is inserted into Eqs. 4a, 4b, 5a, and 5b. The validity of this derivation rests on the assumption that the anion species a is distributed between compartments ′ and ″ according to Donnan's law such that even when a' is not equal to a'' the system is at thermodynamic equilibrium and no net flux of a occurs ("equilibrium exchange").

The pertinent values of V_{max} and $K_{1/2}$, as defined in Table 2, are independent of the concentration of a if $a' = a'' = a$, i.e., when the concentration of a is varied equally on both sides of the membrane and no concentration difference across the membrane exists (first column in the table). When a' is varied at constant a'' (second column in the table) or a'' at constant a' (third column in the table) both V_{max} and $K_{1/2}$ are functions of the (constant) anion concentration at the opposite membrane surface. However, when a' is varied at constant a'' while a'' is very much larger than $qK_{11}/(1+q)$, then $K'_{1/2} = K_{101}/(1+q)$; and when a'' is varied at constant a' while a' is much larger than $K_{101}/(1+q)$, then $K''_{1/2} = qK_{11}/(1+q)$. In other words, when the anion concentration at one membrane surface is made high enough to saturate the transport system, then the V_{max} and $K_{1/2}$ values observed after variations of the anion concentration at the other membrane surface are independent of this constant concentration. For this experimental condition the calculation of the mass law and rate constants defined by the model is particularly convenient.

For chloride transport in human red cells, one of the important predictions of Ping-Pong kinetics could be verified: V_{max} as determined by varying the external Cl^- concentration was found to depend on the level of intracellular Cl^- that is maintained constant while the external Cl^- concentration is varied (Fig. 10; *Gunn* and *Fröhlich* 1979; *Hautmann* and *Schnell* 1985). Corresponding observations were made when the internal Cl^- concentration was varied at a range of fixed external Cl^- concentrations (*Hautmann* and *Schnell* 1985). For infinite trans concentration, it is possible to calculate separately the half saturation constants that pertain to the two sets of experimental conditions. *Gunn* and *Fröhlich* calculated separately $K'_{1/2} = K_{101}/(1+q)$ and $K''_{1/2} = qK_{11}/(1+q)$. The values obtained were about 60 and 3 mmol $liter^{-1}$ for the inner and outer mem-

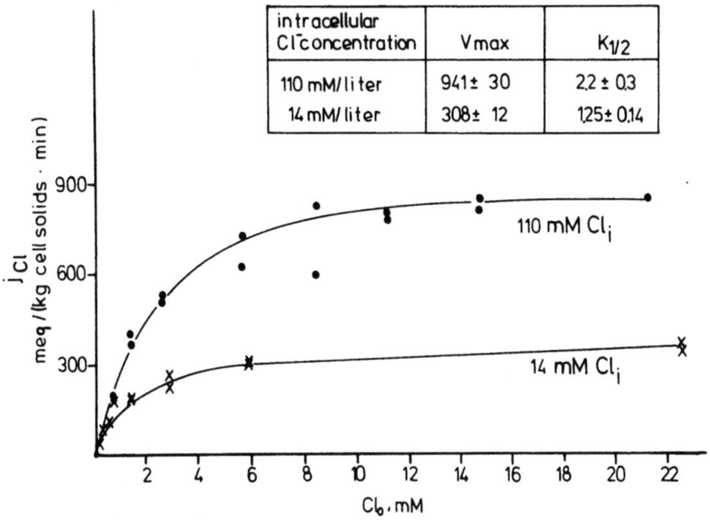

Fig. 10. Chloride equilibrium exchange as measured at the two fixed intracellular Cl^- concentrations represented in the figure and at the extracellular Cl^- concentrations indicated *on the abscissa.* Human red cell ghosts, pH 7.8, 0°C. *Ordinate,* chloride efflux as measured by means of ^{36}Cl. (*Gunn* and *Fröhlich* 1979)

brane surface, respectively (pH 7.8, 0°C). Using the data of *Hautmann* and *Schnell,* one arrives at 14.5 and 4.7 mmol liter^{-1}, respectively (pH 7.3, 0°C). The $K_{1/2}$ values for varying chloride at equal concentrations in internal and external medium were 60 mmol liter^{-1} (*Gunn* and *Fröhlich* 1979) and 21 mmol liter^{-1} (*Hautmann* and *Schnell* 1985).

The analysis of anion transport by measuring equilibrium exchange under the conditions specified in Table 2 neither allows one to calculate q nor the individual rate constants k_{12} and k_{21}. A provisional value of q could, however, be derived from other types of work, suggesting that $K_{101} = K_{11}$, i.e., that the affinity for chloride binding to the transport protein with outward-oriented transfer site is identical to the affinity of the protein with inward-directed transfer site (*Knauf* and *Mann* 1982, 1984b). On this assumption, one can calculate from the expressions for $K'_{1/2}$ and $K''_{1/2}$ a numerical value for $q = k_{12}/k_{21}$. The data of *Gunn* and *Fröhlich* yield $q = 0.05$; those of *Hautmann* and *Schnell* 0.3–0.4. Although the discrepancy is considerable it is clear that in a chloride medium at 0°C most of the transfer sites face toward the cell interior.

Recruitment. Establishment of concentration differences between cell interior and medium is a convenient method of recruiting transport molecules into conformations with inward- or outward-facing transfer sites. For the special case of a single anion species, one obtains for the asymmetry factors

$$\frac{ar + r}{as + s} = \frac{A'K''}{A''K'} = \frac{(1 + \frac{K_{101}}{a'}) \cdot k_{21}}{(1 + \frac{K_{11}}{a''}) \cdot k_{12}} \tag{7}$$

and

$$\frac{r}{s} = \frac{K''K_{101}\, a''}{K'K_{11}\, a'} = \frac{a''K_{101}\, k_{21}}{a'K_{11}\;\, k_{12}} \tag{8}$$

Thus a concentration difference of a across the membrane recruits the transfer sites of the unloaded band 3 protein molecules toward that surface that is in contact with the lower concentration of a. It may be noted that for $K_{101} = K_{11}$, at equal concentrations of a at both membrane surfaces ($a' = a'' = a$) the asymmetry factors $(ar + r)/(as + s)$ and r/s are independent of the absolute value of a (see *Knauf* 1985). However, if K_{101} is unequal to K_{11} the former of the two asymmetry factors, in theory at least, varies with the concentration of a, even though no concentration difference exists (*Passow* et al. 1980; *Passow* and *Fasold* 1980).

Recruitment by establishing anion gradients across the membrane has been extensively used in studies about the chemical or enzymatic modification of band 3 protein where one wishes to discriminate between the actions on conformers with inward- and outward-directed transfer sites. Examples are provided by the work of *Jennings* (1980, 1982b), *Jennings* and *Adams* (1981), *Knauf* (1982), *Wieth* and *Bjerrum* (1982), and others. The results will be discussed below in the context of biochemical studies on band 3-mediated anion transport.

3.1.6.3 Net or Hetero Exchange

The experimental work required to obtain the data necessary for the evaluation of the various rate and equilibrium constants by the methods described in the preceding section is considerable. It is most useful, therefore, that there exists a rather simple method to determine the true mass law constants K_{11} and K_{101} (*Jennings* and *Adams* 1981; *Jennings* 1982b). It is based on the fact (*Jennings* and *Adams* 1981; *Jennings* 1982b) that the exchange of two anion species against one another leads to a recruitment of the transfer sites toward that surface which is in contact with the more slowly penetrating anion species. It involves measuring the half saturation constant of the heteroexchange flux at a fixed and saturating concentration of the faster penetrating anion species at one membrane surface and varying the concentration of the more slowly penetrating species at the other. When measured under this special condition, the half saturation constant is approximately equal to the mass law constant for the binding of the slowly penetrating species to the transfer site when that site is oriented toward the compartment that contains the slowly penetrating anion species. This conclusion can be derived as follows:

The net exchange of anion species *a* against anion species *b* is defined as

$$j_{12}^{net} = k_{12}ar - k_{21}as \tag{9}$$

Combining Eqs. i, ii, and 5 one obtains:

$as = \phi \cdot ar$, where $\phi = K'/K''$

and

$$j_{12}^{net} = \frac{k_{12} - k_{21}\phi}{A' + A''\phi} \cdot \overline{RS} \tag{10}$$

In the situation described above it is assumed that *a* penetrates much more slowly than *b* ($k_{12} \ll k_{15}; k_{21} \ll k_{51}$) and that $a'' = b' = 0$. Inserting this into Eqs. 5a and 5b for the calculation of A', A'', K', and K'', one obtains

$$j_{12}^{net} = \frac{k_{12}\overline{RS}}{1+Q} \cdot \frac{a'}{a' + \dfrac{K_{101}}{1+Q}} \tag{11}$$

where

$$Q = \frac{k_{12}}{k_{51}}\left(1 + \frac{K_{31}}{b''}\right)$$

If it is further stipulated that $b'' \gg K_{31}$, then one arrives at

$$V_{max} = k_{12}\overline{RS} \tag{12a}$$

$$K'_{1/2} = K_{101} \tag{12b}$$

Conversely, if one measures the flux in the opposite direction under the conditions $a' = b'' = 0$ and $b' \gg K_{301}$, then one obtains, by analogous reasoning:

$$V_{max} = k_{21}\overline{RS} \tag{13a}$$

$$K''_{1/2} = K_{11} \tag{13b}$$

Jennings and *Adams* (1981) made use of Eq. 13b for the determination of the affinity of sulfate to the band 3 protein with outward-oriented transfer site. They placed red cells containing chloride at a saturating concentration into media containing sulfate at a range of concentrations and measured sulfate influx. Since sulfate penetrates much more slowly than chloride (i.e., $k_{12} \ll k_{15}; k_{21} \ll k_{51}$), the conditions are fulfilled on which

Eq. 13b rests. Hence the experimentally observed value of 10–20 mmol liter^{-1} for $K'_{1/2}$ should indeed correspond closely to the true mass law constant K_{11} for sulfate binding to the outward-oriented transfer site.

3.1.6.4 Competitive Inhibition by Nonpenetrating Inhibitors

The equations for the action of a nonpenetrating competitive inhibitor b in compartment $'$ on the flux of anion species a can be obtained from Eqs. 4 and 5, if one assumes: $b'' = 0$, $k_{15} = k_{51} = 0$. The equation can then be rewritten in the form:

$$j_{12} = k_{12}\overline{RS}\,\frac{a'}{K_{101}}\cdot\frac{K'_{\mathrm{I}}}{K'_{\mathrm{Iapp}} + b'} \tag{14a}$$

where

$$K'_{\mathrm{Iapp}} = K'_{\mathrm{I}}\,[1 + \frac{a'}{K'^{\infty}_{1/2}}\,(1 + \frac{K''^{\infty}_{1/2}}{a''})] \tag{14b}$$

and $K'_{\mathrm{I}} = K_{301}$; $K'^{\infty}_{1/2} = K_{101}/(1+q)$; $K''^{\infty}_{1/2} = K_{11}\,q/(1+q)$.

The latter two quantities represent the apparent dissociation constants for the substrates at surfaces $'$ and $''$, respectively, as measured at infinitively high concentration of a at the opposite surface (a'' and a' respectively).

If the inhibitor resides in compartment $''$ instead of $'$, then $b'' = 0$. Assuming again $k_{15} = k_{51} = 0$, Eqs. 4 and 5 yield:

$$j_{12} = k_{12}\overline{RS}\,\frac{a''}{qK_{11}}\cdot\frac{K''_{\mathrm{I}}}{K''_{\mathrm{Iapp}} + b''} \tag{15a}$$

where

$$K''_{\mathrm{Iapp}} = K''_{\mathrm{I}}\,[1 + \frac{a''}{K''^{\infty}_{1/2}}\,(1 + \frac{K'^{\infty}_{1/2}}{a'})] \tag{15b}$$

and $K''_{\mathrm{I}} = K_{31}$. The other symbols have the same meaning as above.

Two limiting cases may be considered:

1. The substrate concentration is increased in compartment $'$, which contains the inhibitor at the concentration b', while the substrate concentration in the opposite compartment without inhibitor is maintained constant. Under these conditions, when $a'/K'_{1/2} \gg 1$, then $K'_{\mathrm{Iapp}} \to \infty$ and $j_{12} = k_{12}\overline{RS}/(1+q)$ (see Eqs. 14a, 14b). The competing substrate abolishes inhibition and the flux reaches V_{max}.

2. The substrate concentration is increased in compartment $''$, opposite to the compartment $'$ that contains both inhibitor and substrate at fixed concentrations. If $a'/K'_{1/2} \ll 1$ and $a''/K''_{1/2} \gg 1$ then $K'_{\mathrm{Iapp}} = K'_{\mathrm{I}}$. This means that K_{Iapp} tends to approach K_{I} when the transport protein mole-

cules are recruited by a substrate gradient toward the compartment that contains the inhibitor. This consequence of the Ping-Pong model forms the basis for a simple qualitative test to demonstrate the occurrence of recruitment of a transport protein. When an increase of the substrate gradient between compartment ′ and ″ at constant substrate concentration in compartment ′ increases the inhibition produced by a given inhibitor concentration b', one may suspect that a recruitment of the transport protein molecules to surface ′ is involved.

Equations 14a or 15a and 14b or 15b are special cases of a more general treatment by *Passow* et al. (1980b) and had first been derived in somewhat different form by *Fröhlich* (1982). The latter author applied them to the study of competition between the stilbene disulfonate DNDS and chloride in the red cell. The results will be reviewed in conjunction with the discussion of the stilbene disulfonate binding site.

3.1.6.5 Some Inferences from Work with a Noncompetitive Inhibitor: The Affinities for Substrate Binding to Inner and Outer Surface are Equal, $K_{11} = K_{101}$?

As has been shown above, for the derivation of numerical values from measurements of V_{max} and $K_{1/2}$ (Table 2) of the individual parameter values of Ping-Pong kinetics (q, K_{11}, K_{101}), at least one additional relationship between two of these values needs to be known. Work by *Knauf* et al. (1981) and *Knauf* and *Mann* (1984b) suggests that $K_{11} = K_{101}$.

This suggestion has been derived from studies with niflumic acid, a compound that according to *Cousin* and *Motais* (1979) is a strictly noncompetitive inhibitor. At 150 mmol liter^{-1} Cl^- inside and outside the red cells, the chloride flux was inhibited by 50% at 0.63 μmol liter^{-1} niflumic acid. When the external Cl^- concentration was reduced to 10 mmol liter^{-1} at unchanged internal Cl^- concentration, the niflumic acid concentration required to produce 50% inhibition was decreased by one-half. This effect is related to the recruitment of the transfer sites from inward-facing (at equal Cl^- concentrations inside and outside) to outward-facing (in the presence of the outward-directed Cl^- gradient). This follows from the observation that establishment of $Cl_i^- = Cl_o^-$ = 10 mmol liter^{-1} increased the niflumic acid concentration needed to produce 50% inhibition to the value of 0.63 μmol liter^{-1} originally observed at 150 mmol liter^{-1} Cl^- inside and outside.

Knauf and associates interpret their findings on the assumption that the conformational changes of the transport protein accompanying the recruitment of the transfer site from inward-facing to outward-facing induces an increase of the affinity of the modifier site involved in niflumic acid binding. Thus, niflumic acid can be used to monitor the fraction of band 3 molecules with outward-directed transfer sites.

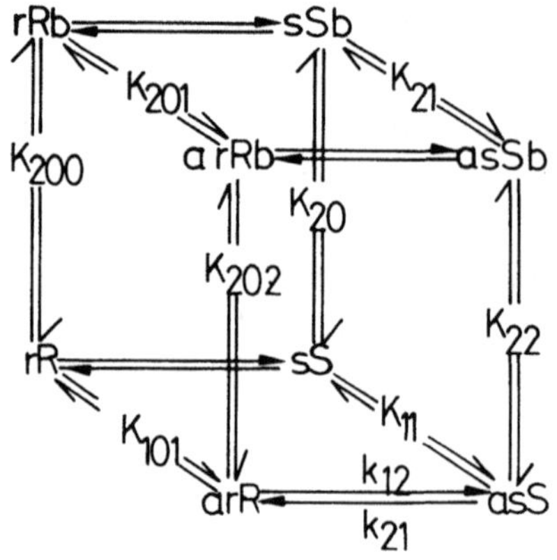

Fig. 11. Noncompetitive inhibition. Reaction diagram showing the binding of the substrate a to the transfer site in conformations r and s, and the inhibitor b to the modifier site in conformations R and S. One should note that the transfer site in r conformation (*arR, arRb, rR, rRb*) faces medium $'$ in s conformation (*asS, asSb, sS, sSb*) medium $''$, while the modifier site faces medium $''$, regardless of whether it is in conformation R (*rR, arR, arRb, rRb*) or S (*sS, asS, asSb, sSb*)

Knauf and his associates (*Knauf* 1981; *Knauf* and *Mann* 1982, 1984b; *Knauf* et al. 1981) have now pointed out that even when the Cl^- concentration is varied on both surfaces equally (i.e., $a' = a''$) Eqs. 6a and 6b still predict changes of the distribution ratio between inward-facing and outward-facing conformers if the affinities K_{101} and K_{11} are different (*Passow* et al. 1980b). The work of *Cousin* and *Motais* shows, however, that the inhibition by niflumic acid is completely independent of the variations of the equal Cl^- concentrations in cells and medium. *Knauf* et al. concluded, therefore, that K_{101} is equal to K_{11}. If this is accepted, one can calculate from the measured values of $K'_{1/2}$ and $K''_{1/2}$ (see Table 2) the individual values of q and $K_{11} = K_{101}$ (see p. 104/105).

It is instructive to consider the reasoning of *Knauf* et al. in some more detail since it illustrates a more general aspect of noncompetitive inhibition in systems with Ping-Pong kinetics (see Fig. 11).

For noncompetitive inhibition, there is no binding of the substrate a to the modifier site (i.e., the site that combines with inhibitor b) in R or S conformation and no binding of the inhibitor b to the transfer site in r or s conformation. Furthermore, b acts only at one surface, e.g., surface $''$, where it reacts with both R and S. This is due to the fact that the modifier site remains exposed to compartment $''$, regardless of whether the transfer site is oriented toward compartment $'$ (r) or $''$ (s). However, r is only accessible to a', s to a''. In terms of the reaction diagram in Fig. 11, this means that *rR, arR, sS,* and *asS* react with b'', while among *rR, rRb, sS, sSb* the forms *rR* and *rRb* react with a', and the forms *sS* and *sSb* with a''. Differences of the affinities of R and S to b'' account for *Knauf's* et al. (1981, 1984) observation that recruitment of the transfer site into r or s form affects the sensitivity of the transport system against inhibition by b.

These considerations may be formulated quantitatively, using the reaction diagram in Fig. 11 and assuming mass law equilibria at the phase boundaries, steady state, and mass conservation.

For the equilibrium exchange flux one obtains:

$$j = \frac{k_{12}\,\overline{RS}}{1+\dfrac{b''}{K_{202}} + q\left(1+\dfrac{b''}{K_{22}}\right)} \cdot \frac{a'a''}{a'a'' + \dfrac{K_{101}\cdot\left(1+\dfrac{b''}{K_{200}}\right)\cdot a'' + qK_{11}\cdot\left(1+\dfrac{b''}{K_{20}}\right)\cdot a'}{1+\dfrac{b''}{K_{202}} + q\left(1+\dfrac{b''}{K_{22}}\right)}} \tag{16}$$

where

$q = k_{12}/k_{21}$

For $b'' = 0$, Eq. 16 yields the uninhibited flux j_{12} $(b'' = 0)$ and becomes identical to Eq. 1. The inhibition is defined as

$I = 1 - [j_{12}/j_{12\,(b\,=\,0)}]$

Insertion of the corresponding expression for j_{12} and $j_{12\,(b''=0)}$ leads to:

$$I = 1 - \frac{a'a''(1+q) + K_{101}a'' + qK_{11}a'}{a'a''\left[1+\dfrac{b''}{K_{202}} + q\left(1+\dfrac{b''}{K_{22}}\right)\right] + K_{101}\left(1+\dfrac{b''}{K_{200}}\right)a'' + qK_{11}\left(1+\dfrac{b''}{K_{20}}\right)a'} \tag{17}$$

from which, by simple algebraic rearrangement, the inhibitor concentration b''_{I} can be calculated that produces the inhibition I. *Knauf* et al. (1984) determined b''_{I} in the presence and absence of a chloride gradient across the membrane. For these two cases one obtains for b''_{I} the following expressions:

$$\underset{a' \gg a''}{b''_{\mathrm{I}}} = \frac{I}{1-I} \cdot \frac{a''(1+q) + qK_{11}}{a''\left(\dfrac{1}{K_{202}} + \dfrac{q}{K_{22}}\right) + q\dfrac{K_{11}}{K_{20}}} \tag{18a}$$

and

$$\underset{a' = a''}{b''_{\mathrm{I}}} = \frac{I}{1-I} \cdot \frac{a''(1+q) + K_{101} + qK_{11}}{a''\left(\dfrac{1}{K_{202}} + \dfrac{q}{K_{22}}\right) + q\dfrac{K_{11}}{K_{20}} + \dfrac{K_{101}}{K_{200}}} \tag{18b}$$

For noncompetitive inhibition, it is plausible to assume that the affinity of the modifier site in R or S conformation is independent of substrate binding to r or s, respectively. Hence $K_{200} = K_{202}$ and $K_{20} = K_{22}$.

Experimentally, it was observed that the niflumic acid concentration required to produce 50% inhibition in the presence of the gradient was lower than the concentration required to produce the same inhibition in the absence of a gradient. In terms of the above equations (assuming $''$ to refer to the extracellular compartment) this implies $K_{20} > K_{200}$. The affinity of the modifier site toward the inhibitor is larger when the transfer site is facing outward than when it faces inward. The inhibitor "senses" the orientation of the transfer site.

Cousin and *Motais* (1979) found no dependence of the inhibition by niflumic acid on substrate concentration. In their work they varied the chloride concentration in cells and medium equally and in the absence of a chloride gradient across the membrane. Hence $a' = a''$. For this condition, Eq. 17 can be transformed into:

$$I = 1 - \frac{(1+q)}{1+\frac{b''}{K_{202}}+q(1+\frac{b''}{K_{22}})} \cdot \frac{a+\frac{K_{101}+qK_{11}}{1+q}}{a+\frac{K_{101}\cdot(1+\frac{b''}{K_{200}})+qK_{11}\cdot(1+\frac{b''}{K_{20}})}{1+\frac{b''}{K_{202}}+q\,(1+\frac{b''}{K_{22}})}}$$

According to this equation, I can only be independent of substrate concentration if

$$\frac{K_{101}+qK_{11}}{1+q} = \frac{K_{101}(1+\frac{b''}{K_{200}})+qK_{11}\,(1+\frac{b''}{K_{20}})}{1+\frac{b''}{K_{202}}+q\,(1+\frac{b''}{K_{22}})}$$

If $K_{20} = K_{22} > K_{200} = K_{202}$, then this equation can only be true if $K_{101} = K_{11}$.

It is interesting to note that, according to the theoretical interpretation presented above, the independence of the inhibition of the substrate concentration $a' = a'' = a$ is only a consequence of the accident that in the particular case of the band 3 protein, inward- and outward-oriented transfer sites have equal affinities for the substrate. If this is not the case (i.e., when $K_{101} \neq K_{11}$), then the transfer site will be recruited, even at equal substrate concentrations, on both surfaces of the membrane (*Passow* and *Fasold* 1980; *Passow* et al. 1980b). Under this condition, the accompanying conformational changes of the modifier site (i.e., the site which combines with the inhibitor) would make the inhibition dependent on the substrate concentration, contrary to what one would intuitively expect for a noncompetitive inhibition.

Previous work on the inhibition of anion transport by dinitrophenylation of a specific lysine residue (see p. 148ff.) has been interpreted on the assumption that the asymmetrical organisation of the transport protein in the membrane precludes the possibility that $K_{101} = K_{11}$. Changes of the inhibition brought about by equal variations of the substrate concentration on both surfaces of the membrane ($a' = a'' = a$) under otherwise identical conditions were interpreted, therefore, as indicative for recruitment of the substrate binding site (*Passow* et al. 1980a,b; *Passow* and *Fasold* 1980). This interpretation is not compatible with the interpretation of the experiments with niflumic acid. Further work is needed to resolve which of the conflicting interpretations is correct.

The discussion presented above is not only important in the context of the work of *Knauf* and his associates. It is generally necessary to realize that in transport systems with Ping-Pong kinetics noncompetitive inhibition will usually depend on substrate concentration and thus exhibit a feature that is held to be a typical indicator for competitive inhibition.

3.1.6.6 Temperature Dependence of Anion Transport

The apparent activation enthalpy of anion exchange is rather high (see below). Each one of the four coefficients required to describe the transport processes in terms of the model in Figs. 5–7 could, in principle at least, vary with temperature and hence contribute to the observed enthalpy change. However, the apparent $K_{1/2}$ value of anion transport does not seem to change much with temperature. For example, *Brahm* (1977) measured the dependence on Cl^- of chloride equilibrium exchange at nearly equal chloride concentrations inside and outside the red cells and obtained for $K_{1/2} = (qK_{11}+K_{101})/(1+q)$ values that increased from 28 mmol liter^{-1} at 0°C to 65 mmol liter^{-1} at 38°C. This would correspond to an insignificant contribution of at most 2–3 kcal mol^{-1} to the apparent activation enthalpy. Further indications to the same effect have been obtained for sulfate ions (*Gunn* 1978; *Ruffing* and *Passow*, unpublished); In addition, *Fröhlich* (1982) discusses some evidence in favor of the assumption that the enthalpy change of the true affinity of the anions at least to the outward-facing conformer of the transport protein is quite small. Thus, the essential contributions to the activation enthalpy should come from the temperature dependence of V_{max}. Most of the available data are anyway derived from measurements of anion equilibrium exchange at nearly saturating substrate concentrations where, according to Table 2

$$V_{max} = \frac{k_{12}\overline{RS}}{1+\dfrac{k_{12}}{k_{21}}} = \frac{k_{12}^{o}\exp\dfrac{-E_{12}}{RT}}{1+\dfrac{k_{12}^{o}}{k_{21}^{o}}\exp\dfrac{(E_{21}-E_{12})}{RT}}\,\overline{RS} \tag{19}$$

E_{12} and E_{21} represent, respectively, the activation enthalpy pertaining to k_{12} and k_{21}, while k_{12}^{o} and k_{21}^{o} represent corresponding temperature-independent coefficients.

There are two limiting cases. If $k_{21} \gg k_{12}$, then $V_{max} = k_{12}\overline{RS} = k_{12}^{o}\exp(-E_{12}/RT)$; if $k_{21} \ll k_{12}$, then $V_{max} = k_{21}\overline{RS} = k_{21}^{o}\exp(-E_{21}/RT)$. Thus, sufficiently large changes of the equilibrium between inward- and outward-facing conformers with temperature (i.e., of the ratio k_{12}/k_{21}) should produce a break in the Arrhenius plot.

For slowly penetrating anion species such as SO_4^{2-}, the Arrhenius plot of the self-exchange flux yields a straight line over the whole range between 0°C and 37°C with apparent activation enthalpies of about 30–35 kcal/mol (*Passow* 1969, and many others, cited by *Knauf* 1979). The absence of a break suggests that over the whole temperature range $k_{12}/k_{21} \ll 1.0$ or that the temperature dependence is dominated by k_{12}^{o} and the corresponding activation enthalpy E_{12}, or by k_{21}^{o} and the corresponding activation enthalpy E_{21}, or that $E_{12} = E_{21}$.

The various possibilities can be tested experimentally. For this purpose it is necessary to recruit the transfer sites of all transport molecules either inward or outward and to measure the temperature dependence of transport under these two conditions. For SO_4^{2-} transport this can be achieved by measuring the temperature coefficient of net exchange of SO_4^{2-} against the more rapidly penetrating Cl^-, when (a) SO_4^{2-} is inside and Cl^- outside and (b) when Cl^- is inside and SO_4^{2-} outside (see p. 106ff.). Some preliminary experiments by *Legrum* and *Passow* at 20°C and 30°C indicate that in human red cells the apparent activation enthalpies of the two quantities are virtually indistinguishable. This implies that the distribution between sulfate-loaded transport protein molecules with inward-facing and outward-facing transfer sites is independent of temperature (i.e., that the reaction enthalpy associated with reorientation is zero) and that the temperature dependence of the penetration rate is essentially due to the conformational transition of the loaded forms of the transfer site from facing one membrane surface to facing the other.

For the rapidly penetrating Cl^- and Br^-, Arrhenius plots of the rate of equilibrium exchange show breaks at 15°C and 25°C, respectively. Below the break, the apparent activation enthalpy is about 30 kcal mol^{-1}, above about 20 kcal mol^{-1} (*Brahm* 1977). At the break temperatures the turnover numbers of Cl^- and Br^- transport are about equal (4×10^3 ions band 3^{-1} s^{-1}), and hence *Brahm* suggested that when this turnover number is reached a transition from one rate-limiting step to another takes place. In terms of the Ping-Pong model this would imply that at all temperatures $k_{12}/k_{21} \ll 1$ and that the break is due to a change of k_{12}, which in the present context refers to the transition from inside to outside. However, it would also be consistent with the Ping-Pong model of anion transport to assume that in the case of the anion species mentioned E_{12} is unequal to E_{21}. The enthalpy change $\Delta H = E_{21} - E_{12}$ of about 10 kcal mol^{-1} would be associated with the change with temperature of the equilibrium between band 3 molecules with inward- and outward-facing transfer sites. If this interpretation were correct, one would have to assume that with increasing temperature the halide-loaded transfer sites are recruited from predominantly inward facing to predominantly outward facing.

Eidelman and *Cabantchick* (1983a) have measured the *net* exchange of a fluorescent anion (NBD-taurine) in red cells against Cl^-, Br^-, or SO_4^{2-} in the medium. They incorporated the dye at low concentration into red cells that had been equilibrated at a high concentration of either one of the three anion species mentioned. The dye-loaded red cells were then washed and resuspended in media of the respective anion composition to follow the net efflux of the dye at a range of temperatures. They found that in SO_4^{2-} medium the Arrhenius plot yielded a straight line with an apparent activation enthalpy E_A = 30 kcal mol^{-1}. In Cl^- and Br^- medium, deviations were found which could be described by two different activation enthalpies below and above certain critical temperatures (17° ± 3°C and 29° ± 5°C for Cl^- and Br^-, respectively) which were similar to those observed by *Brahm* (1977) in his studies of the equilibrium exchange of Cl^- and Br^-. Perhaps the findings of *Eidelman* and *Cabantchik* can be interpreted on the assumption that the anion species that predominates in cells and medium determines the distribution ratio between band 3 molecules with inward- and outward-facing transfer sites and that different temperature dependences of these ratios are responsible for the differences of the temperature dependences of NBD-taurine transport in sulfate and halide media. Since NBD-taurine, in contrast to sulfate and the halides, was present in a nonsaturating concentration, changes of temperature could possibly also affect NBD-taurine binding to the transport protein, which would express itself as an additional effect on the temperature dependence of NBD-taurine transport. Further work will be needed to clarify whether the additonal effect would be as small as for the inorganic anions or if it plays a predominant role for the interpretation of the data as suggested by *Eidelman* and *Cabantchik*.

The high absolute values of E_{12} and E_{21} still require an explanation and it may be mentioned that *Ross* and *McConnell* (1978) proposed that the band 3 protein may exist in the red cell membrane in transporting and nontransporting forms and that the equilibrium between the two forms changes with temperature. The enthalpy change of this reaction could contribute to the apparent activation enthalpy of anion transport derived from the Arrhenius plot.

3.1.6.7 *Pressure Dependence*

The temperature dependence of transport (at constant pressure) is usually expressed in terms of the first derivative of the measured rate konstant $^\circ k$ with respect to $1/T$, the inverse of the absolute temperature:

$$R \cdot \frac{\partial ln^\circ k}{\partial 1/T} = E_A \quad \text{or} \quad RT \frac{\partial ln^\circ k}{\partial T} = -\frac{E_A}{T}$$

These equations define the activation enthalpy discussed in the preceding section.

Similarly, a coefficient describing the pressure dependence (at constant temperature) can be defined by the formation of the first derivative of the rate constant with respect to pressure P:

$$RT \cdot \frac{\partial ln^\circ k}{\partial P} = \Delta V^*$$

where the coefficient ΔV^* is called the activation volume.

Canfield and *Macey* (1984) have studied the pressure dependence of sulfate equilibrium exchange in intact human red cells. At 30°C, over the pressure range 0.1–83 MPa (equivalent to 15–12 000 lb/in^2) they find increasing inhibition of transport with increasing pressure. The changes were completely reversible and followed the straight-line relationship predicted by the above equation when $ln\ ^\circ k_{SO_4}$ was plotted against P. From the slope, $\Delta V^* = 150\ cm^3\ mol^{-1}$ was calculated.

The authors pointed out that this activation volume is far above the range reported for enzyme reactions in aqueous media (–30 to +30 cm^3 mol^{-1}) and even much higher than the value reported for the Na^+–K^+ pump in red cells of 65 $cm^3\ mol^{-1}$. They discuss their result in terms of volume changes that were possibly associated with anion binding and the conformational changes of the transport protein during anion translocation across the cell membrane. By partial differentiation with respect to P of the anion flux as described by Ping-Pong kinetics (see Eq. 1, with $a' = a'' = a$), they arrive at the following expression:

$$\frac{\partial ln j_{12}}{\partial P} = -\frac{K_{1/2}}{K_{1/2}+a} \cdot \frac{\partial ln K_{1/2}}{\partial P} + \frac{\partial ln\ V_{max}}{\partial P}$$

where

$$K_{1/2} = \frac{K_{101} + qK_{11}}{1+q} \quad \text{and} \quad V_{max} = \frac{k_{12} \cdot k_{21}}{k_{12} + k_{21}} \quad \overline{RS}$$

In this equation, $\partial ln\ j_{12}/\partial P$ corresponds to the activation volume $\Delta\ V^*$ and $\partial ln\ K_{1/2}/\partial P$ to $\Delta\ V_b$, the volume change pertaining to anion binding to the transfer site. A careful discussion of their results in the light of this equation led the authors to conclude that ΔV_b is much smaller than $\Delta\ V^*$ and hence that the latter largely reflects the volume changes $\partial ln\ V_{max}/\partial P$ that refer to the conformational changes of the transport protein during anion translocation. This conclusion is well supported by many other findings on the kinetics of band 3 protein-mediated anion transport and will reappear as a leitmotiv throughout the rest of the review.

3.1.6.8 Alternatives to Ping-Pong Kinetics

It has been repeatedly stated that in systems without modifier sites Ping-Pong kinetics lead to equations that are formally identical to the Michaelis-Menten equation. The saturation kinetics predicted should show a Hill coefficient of 1.0. This has actually been observed in many instances, most recently by *Hautmann* and *Schnell* (1985), who were able to extend their measurements to anion concentrations below 1 mmol liter^{-1}. This does not exclude the possibility that below the range were Ping-Pong kinetics have been established one or several additional anions are bound and that such anion binding could be a prerequisite for the transport process to take place. Thus, it has been speculated that anion translocation across the membrane can only take place when two anions are bound to the transport protein at the opposite membrane surfaces and that during the translocation step these anions change place. This would imply that there exists an obligatory coupling of the movement of two anions with the transport-related conformational change of each band 3 protein molecule. Such a mechanism would be entirely different from the Ping-Pong mechanism where anion transport is the result of the independent transfer of single anions by each transport molecule and where the 1:1 stoichiometry of the exchange follows from the necessity to maintain electroneutrality in the bulk solutions on the two surfaces of a cation-impermeable membrane.

Attempts to demonstrate the occurrence of coupled anion movements have failed so far (for review, see *Knauf* 1979; *Gunn* and *Fröhlich* 1982). It should be noted, however, that the mathematical formalism of Ping-Pong kinetics would still apply at anion concentrations above the range at which one of the two sites is occupied by the more firmly bound anion. Moreover, a report by *Salhany* and *Rauhenbühler* (1983) suggests deviations from Ping-Pong kinetics which do not seem to fit into the concepts developed above and which should be studied further.

3.1.7 Modification of Substrate Transport by Substrate Binding to Modifier Sites

3.1.7.1 Discussion in Terms of Allosteric Interactions Between Transfer and Modifier Site (Dalmark 1975)

As has already been pointed out (p. 90), the relationship between equilibrium exchange and substrate concentration passes through a maximum. This has been interpreted as a superimposition of saturation of the "transfer site" that is involved in anion translocation and an inhibition by saturation of a "modifier site" (*Dalmark* 1975; see Table 1). For most inorganic anion species studied, the apparent dissociation constants for the modifier site vary roughly parallel with the apparent dissociation constants for the transfer site and are about four to eight times higher than the latter. Bicarbonate and thiocyanate constitute exceptions. The latter shows a particularly low value for the apparent dissociation constant of the transport site and an immeasureably high value for the modifier site (*Dissing* et al. 1981).

An equation that describes the effects of anion binding to the modifier site can be derived from the reaction network in Fig. 9a and Eq. 2 (*Dalmark* 1975). *Dalmark's* derivation involves two simplifying assumptions: (1) in inward- and outward-facing conformers of the transport protein, the respective affinities of transfer and modifier site are equal ($K_{100} = K_{10}$; $K_{101} = K_{11}$; $K_{102} = K_{12}$; $K_{103} = K_{13}$) and (2) anion binding to the modifier site does not change the affinity for anion binding to the transfer site and vice versa ($K_{100} = K_{102}$; $K_{101} = K_{103}$). The ensuing equation reads (*Dalmark* 1975):

$$j_{12} = k_{12} \cdot \overline{RS} \cdot (1+q)^{-1} \cdot (1+K_{101}/a)^{-1} \cdot (1+a/K_{100})^{-1} \qquad (20)$$

where K_{100} and K_{101} are the dissociation constants for modifier and transfer site, respectively, and $q = k_{12}/k_{21}$. It should be noted that allosteric effects on the modifier site (characterized by the differences between K_{100} and K_{102} or K_{10} and K_{12}) due to increasing occupancy of the transfer site should be difficult to detect since K_{100}/K_{101} is approximately 5 (Table 1) and hence in most band 3 molecules the transfer site will be occupied before the occupancy of the modifier site becomes measurable. Further justification for the simplifying assumptions comes from *Knauf* and *Mann's* (1984b) work with niflumic acid. Although the agent is capable of sensing the orientation, the fractional inhibition that it produces is independent of the degree of saturation of transfer *and* modifier site, indicating that chloride binding has no effect on the orientation of the transport molecule. This would only be the case if $K_{101} = K_{11}$ *and* if the orientation of the modifier site did not change upon increasing occupancy with Cl^-.

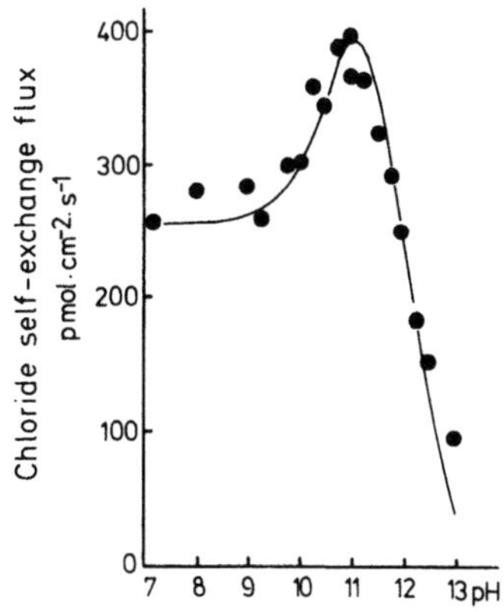

Fig. 12. Dependence of chloride self-exchange on extracellular pH. The intracellular pH was 7.2; the intra- and extracellular Cl^- concentrations were 330 mmol liter^{-1}. The curve was calculated on the assumption that it represents the superimposition of two processes with p*K*s of 10.7 and 12.0. The former process accelerates; the latter inhibits the Cl^- self-exchange. The former process can be inhibited by NAP-taurine and is attributed to an outward-facing substrate-binding modifier site (presumably including an arginine residue) that is different from *Dalmark's* (1975) modifier site. The latter process involves an outward-facing arginine residue which is presumably identical to r_2 in Figs. 16 and 20 or to arg *b* in Table 4. (*Wieth* and *Bjerrum* 1982)

Attempts to find out whether Dalmark's modifier site faces toward the cytosol or the medium have resulted in a number of studies which revealed the existence of two instead of one halide-binding modifier site. One of them faces inward, the other outward. Dalmark's site is identical with the former.

The evidence for this conclusion is as follows: *Knauf* and *Mann* (1984a) varied the extracellular Cl^- concentration from 150 to 600 mmol liter^{-1} at a fixed intracellular Cl^- concentration of 600 mmol liter^{-1} and found no effect on the rate of Cl^- self-exchange. When they varied the internal Cl^- concentration over the same range at a fixed extracellular concentration of 600 mmol liter^{-1}, they observed inhibition. This inhibition was not significantly different from the inhibition produced when intra- and extracellular Cl^- concentrations were varied equally on both surfaces. These observations are in full agreement with essentially similar work of *Gunn* and *Milanick* (1982, 1983). They show clearly the involvement of an inhibitory modifier site at the inner membrane surface with properties that account for the original observations of Dalmark.

The occurrence in the outer cell surface of yet another modifier site is inferred from the work of *Wieth* and *Bjerrum* (1982) and *Knauf* (1985). When the dependence of Cl^- equilibrium exchange on external pH is measured at constant internal pH, a site can be titrated whose deprotonation (p*K* = 10.7) is associated with an enhancement of Cl^- transport (Fig. 12). This site is involved in the binding of NAP-taurine, a sulfonic acid that is accepted as a substrate by the transport system and thus is capable of producing competitive inhibition. At the same time the agent causes inhibition at an additional site which is probably identical to the site that deprotonates with the p*K* of 10.7 (*Wieth* and *Bjerrum* 1982). NAP-taurine (*Knauf* et al. 1978a, 1980; *Knauf* 1982) and a related compound, NIP-taurine (*Knauf* 1984, personal communication), compete with Cl^- for binding to this site, indicating that it is a substrate-binding site. The apparent affinity of Cl^- for this site is similar to that for the intracellular Cl^--binding modifier site (i.e., Dalmark's site). The extracellular site, in contrast to Dalmark's site, responds to variations of the Cl^- concentration gradient across the membrane and hence is allosterically linked to the recruitment of the transfer site. Its functional significance is still obscure.

3.1.7.2 Discussion of Substrate Self-Inhibition in Terms of Substrate Binding to Approach Sites in an Anion Channel with Variable-Energy Barriers (Tanford 1985)

Dalmark's description of the self-inhibition of anion exchange stipulates the existence of a modifier site whose functional significance for the transport process remains unexplained. Recently, *Tanford* (1985) has reinterpreted the self-inhibition by a model which, in the present context, is best described as an expansion of *Läuger's* model of a channel with variable-energy barriers (*Läuger* 1980, 1984). *Tanford's* model assigns a specific function to the sites whose occupancy by the substrate is responsible for self-inhibition.

In accord with existing views, *Tanford* argues that the conformational change that brings about the transport cannot move the anion over a distance as large as the thickness of the lipid bilayer. For this reason he agrees with the idea that the anion-binding site between the rate-limiting alternating energy barriers is located within the protein molecule some distance away from one or both membrane surfaces and that this site is connected to the internal and external solutions by narrow channels. The occupancy of "approach sites" in these connecting channels would determine the probability of the combination of a diffusing anion with the alternating access site or the release of a bound anion from this site into the channel. At low concentrations, the approach sites would direct the penetrating anions toward or away from the alternating access gate and thus facilitate transport. At high substrate concentrations, the approach sites would be occupied by the substrate anions. This would make it impossible for the alternating access gate to release an anion into the adjacent medium and thus lead to self-inhibition (Fig. 6).

An equation for the relationship between equilibrium exchange and substrate concentration predicted by this model has been derived by *Tanford.* It is based on several simplifying assumptions: (1) Each one of the two access channels connecting the alternating access site to cell interior and the medium contains only one approach site; (2) each of the two approach sites is in rapid equilibrium with the adjacent bulk solutions; and (3) there are no interactions between the approach sites and the alternating site. The latter assumption is equivalent to postulating that the equilibrium constants describing anion distribution between the two bulk phases and the corresponding approach sites, and the rate constants for the alternations of the energy barriers of the alternating access site are independent of the occupancy of the approach sites.

On these assumptions, for the unidirectional flux j_{12} the expression:

$$j_{12} = \frac{C \cdot a}{1 + A \cdot a + B \cdot a^2}$$ can be derived.

Here A, B, and C are constants and a is the substrate concentration. This equation is formally identical to the Dalmark equation (Eq. 20), as may be seen after suitable algebraic rearangement. However, the constants have a different physical meaning; they represent rather complex expressions which depend on the occupancy of the various sites and hence on affinity and rate constants. The observation that self-inhibiton of Cl^- exchange occurs exclusively from the inside surface of the membrane would imply for the present model that there exists only one approach site inside the inward-directed approach channel. For this situation, the mathematical structure of the above equation remains unaltered, although the constants A, B, and C no longer include the terms that refer to external approach sites.

It is important to note that Tanford's model predicts self-inhibition only if the rate of exchange of an anion between the approach sites and the alternating site is of the same order of magnitude as the rates of alternation of the heights of the energy barrier at the alternating access site. This would seem to be the case for all anion species studied so far with the exception of thiocyanate (Table 1).

3.1.8 pH Dependence of Monovalent Anion Transport

The pH range over which anion equilibrium exchange can be measured is limited to $5 \leqslant \text{pH} \leqslant 9$. This is predominantly due to effects on the proteins that constitute the cytoskeleton, as has been convincingly documented for the case of low pH. When the pH falls below 5 and thus below the isoelectric point of band 3 protein (*Schubert* and *Domning* 1978) and spectrin (*Wieth* et al. 1980), an aggregation of the intramembrane particles occurs (*Bjerrum* et al. 1980). This is accompanied by a gradual increase of the permeability for small hydrophilic ions and molecules (*Gunn* et al. 1973; *Wieth* et al. 1980). Although these changes are almost completely reversible after retitration to pH 7.2 and incubation at 37°C, they set a lower limit to the study of anion equilibrium exchange at equal pH outside and inside the red cells.

However, titration over a wider range of external pH values can be performed when the intracellular pH is maintained at a constant value within the pH limits where the changes described above do not take place (*Wieth* and *Bjerrum* 1982). Three dissociating groups can thus be identified with p*K*s of about 5.2, 11, and 12. Protonation of the groups with p*K* 5.2 and deprotonation of the groups with p*K* 12 leads to an inhibition of transport, while deprotonation of the group with p*K* 11 results in an activation. The lowest p*K* value can be attributed to a COOH group, the two higher values to arginyl residues. The identity of the groups responsible for the p*K*s of 5.1 and 12.0 has been further established by chemical modification experiments which will be described in a later section.

The arginyl residue that titrates at pH 12 (see Fig. 12) is most likely located in close proximity to the substrate binding site that is involved in anion translocation. Its properties have been studied in much detail. The p*K* value decreases with decreasing Cl^- concentration in the medium, which has been attributed to a change of the local pH in the vicinity of the arginyl residue due to an increase of the negative surface potential (see Sect. 4.3.2).

The outward-facing arginyl residue that titrates at pH 11 is a constituent of one of the substrate-binding modifier sites discussed above (*Wieth* and *Bjerrum* 1982).

Finally, titrating simultaneously the inner and outer surface within the pH limits described above showed an inhibition of equilibrium exchange at pH values below 7.1. This seems to reflect the minor contribution of titratable groups that are in contact with the outer cell surface and the major contribution of a group that is exposed at the inner cell surface (*Wieth* et al. 1980).

There are some conflicting reports on the role of Cl^- in the titration of the intracellular proton-binding site (discussed by *Wieth* et al. 1980). Earlier work by *Dalmark* suggested that the inhibition of the Cl^- transport by H^+ is not affected by chloride binding. In contrast, *Gunn* and *Milanick* (1982, 1983) report that at constant external pH and Cl^- concentration the inhibition not only depends on internal H^+ but also on internal Cl^-. When the internal Cl^- concentration is varied at several fixed internal pH values between 5.7 and 7.8, the V_{max} values are independent of internal pH. However, when the internal pH is varied at a range of constant internal Cl^- concentrations, with increasing internal Cl^- concentration level, the inhibitory effect of H^+ is decreased. When the pH limits indicated above are exceeded, noncompetitive components of the effects of internal protons become apparent (*Gunn* and *Milanick* 1983).

Milanick and *Gunn* (1982a) studied the interactions between binding of extracellular H^+ and Cl^-. They tried to determine the effect of proton binding to the transport protein under conditions where the transfer and modifier site are either unloaded or loaded with one or possibly two Cl^- ions. They found that reaching the peak flux at which activation by substrate saturation of the transfer site is balanced by substrate saturation of the inhibitory modifier site requires a Cl^- concentration of 100 mmol $liter^{-1}$ at pH 6.8, but only 10 mmol $liter^{-1}$ at pH 4.1. Apparently the affinity of an outward-facing inhibitory modifier site to Cl^- is greatly increased by proton binding.

3.2 Divalent Anion Transport

The kinetics of divalent anion transport resemble in many respects the kinetics of monovalent anion transport. Equilibrium exchange shows saturation kinetics with self-inhibition at high substrate concentration (*Passow* et al. 1977; *Schnell* et al. 1977; *Barzilay* and *Cabantchik* 1979b). The activation enthalpy has the same unusually high value as for monovalent anions, and monovalent anions produce a competitive inhibition of divalent anion exchange (*Schnell* et al. 1977). Finally, a large number of inhibitors produce the same fractional inhibition of Cl^- and SO_4^{2-} transport (*Ku* et al. 1979). Nevertheless, there exist two important differences between monovalent and divalent anion equilibrium exchange: (1) The rate of transport for the divalent anions is many orders of magnitude lower than for the monovalent anions. (2) Monovalent anion transport increases with increasing pH until a plateau or a flat maximum is reached above pH 7.0 (*Gunn* et al. 1973; *Funder* and *Wieth* 1976, Fig. 13a), while divalent anion transport passes through a pronounced maximum around pH 6.3–6.5 (*Schnell* et al. 1977, Fig. 13b)[7].

The differences of pH dependence have been related to the observation that divalent anions penetrate across the intact red cell membrane together with a proton (*Jennings* 1976). Thus, when Cl^--containing red cells are suspended in a (CO_2-free) SO_4^{2-} medium, each mole of Cl^- that leaves the cells is replaced by 1 mol H^+ plus 1 mol SO_4^{2-}. This gives rise to an easily measurable increase of the pH in the external medium. Correspondingly, a decrease is seen when SO_4^{2-}-containing red cells are suspended in a Cl^- medium.

The first explanation of the differences between monovalent and divalent anion transport was provided by *Gunn's* model of a titratable carrier (*Gunn* 1973). It was suggested that the transport system could exist in two interconvertible forms: one doubly protonated, the other singly pro-

7 A third difference seems to exist with respect to concentration dependence. For Cl^- transport, down to less than 1 mmol $liter^{-1}$, the relationship between flux and concentration yields a Hill coefficient close to 1.0 (*Hautmann* and *Schnell* 1985). Under certain experimental conditions, according to *Schnell* and *Besl* (1984), for sulfate and phosphate transport a Hill coefficient close to 2.0 is found

tonated; the former combining with divalent anion species, the latter with a monovalent anion species (see *Gunn* 1978, 1979).

In such a system increasing the pH increases the singly protonated form of the carrier and enhances the rate of monovalent anion transport. The concomitant decrease of the divalent forms should decrease the rate of divalent anion transport. A quantitative evaluation of existing data on the pH dependence of SO_4^{2-} equilibrium exchange above the pH maximum yielded an excellent agreement with these predictions (*Passow* and *Wood* 1974). However, an investigation over a much wider pH-range showed discrepancies (*Schnell* et al. 1977). The model predicted that the V_{max} for SO_4^{2-} should be independent of pH and that the pH dependence should essentially be accountable for by variations of the apparent K_m value for substrate binding. It was pointed out by *Knauf* in 1979 that the experimental results of *Schnell* and his associates showed that this prediction of the titratable carrier theory could not be confirmed: both V_{max} and the $K_{1/2}$ value for SO_4^{2-} transport were functions of pH. It was proposed, therefore, by *Legrum* et al. (1980) that the binding of a proton was not the prerequisite of divalent anion binding as postulated in the original hypothesis of *Gunn*. It was shown that *Schnell's* data could be interpreted at least qualitatively if one assumes that sulfate binding may take place regardless of whether or not a proton was previously bound, but that the translocation of the bound SO_4^{2-} ion across the membrane could only take place when a proton was also bound (Fig. 13). This hypothesis was independently developed by *Milanick* (1980) and *Milanick* and *Gunn* (1982a) and studied in quantitative detail by determining the influence of H^+ and SO_4^{2-} on Cl^- transport at 0°C. It was observed that protons and sulfate were both able to inhibit Cl^- transport and that the effects were cumulative rather than mutually exclusive. It was further shown that SO_4^{2-} could inhibit Cl^- transport by binding when no proton is bound and that SO_4^{2-} and proton binding facilitated each other, each reducing the apparent K_I for the other to about 10% of the original value. Under the experimental conditions described, in contrast to Cl^- transport, the rate of the SO_4^{2-} transport is essentially limited by the supply of protons. This would explain not only why SO_4^{2-} transport decreases with increasing pH while Cl^- transport remains constant, but also, to a certain extent, why the absolute values of the exchange rates are much lower for SO_4^{2-} than for Cl^-.

Although the observations described above remove an obvious objection against *Gunn's* model of a titratable carrier, it remains to be shown that the proton whose binding to the band 3 protein activates sulfate transport is the same as that responsible for the inhibition of chloride transport. The demonstration of a causal relationship between the inverse effects of H^+ on Cl^- and SO_4^{2-} equilibrium exchange encounters the dif-

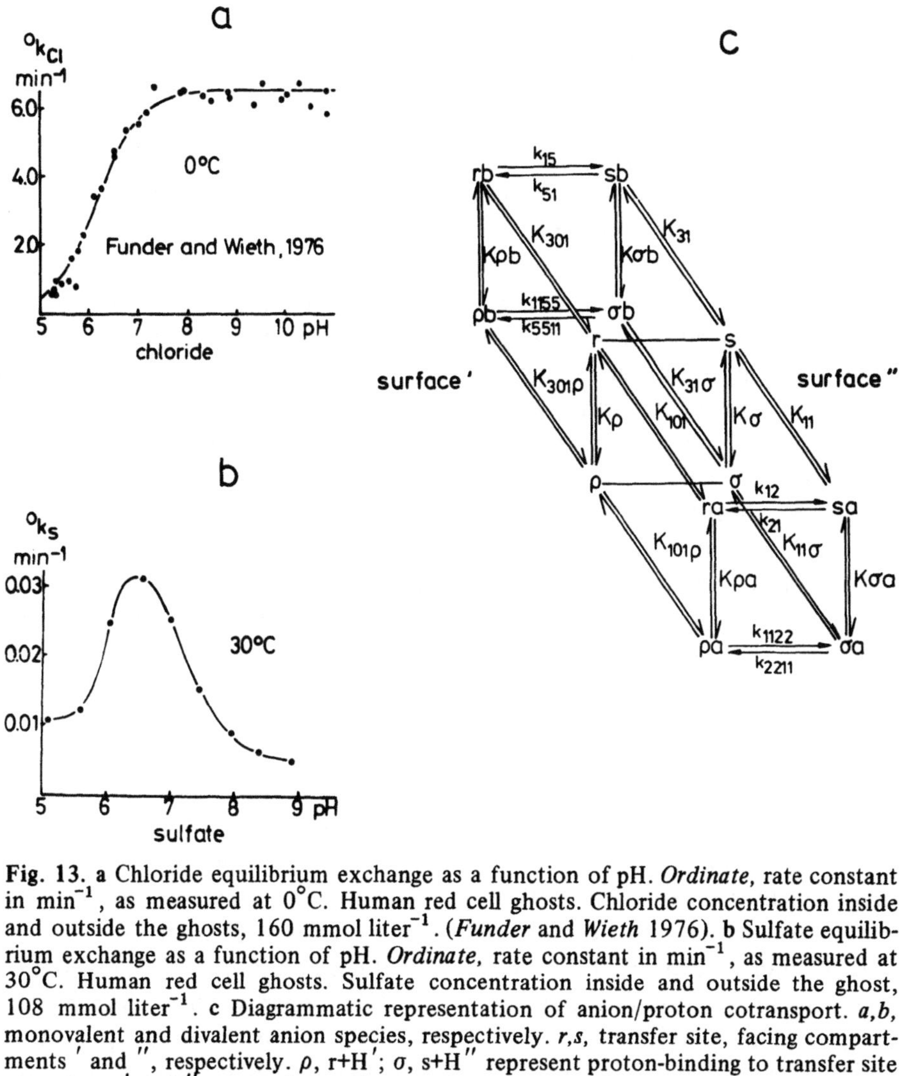

Fig. 13. a Chloride equilibrium exchange as a function of pH. *Ordinate*, rate constant in min^{-1}, as measured at 0°C. Human red cell ghosts. Chloride concentration inside and outside the ghosts, 160 mmol $liter^{-1}$. (*Funder* and *Wieth* 1976). b Sulfate equilibrium exchange as a function of pH. *Ordinate*, rate constant in min^{-1}, as measured at 30°C. Human red cell ghosts. Sulfate concentration inside and outside the ghost, 108 mmol $liter^{-1}$. c Diagrammatic representation of anion/proton cotransport. *a,b*, monovalent and divalent anion species, respectively. *r,s*, transfer site, facing compartments $'$ and $''$, respectively. ρ, r+H$'$; σ, s+H$''$ represent proton-binding to transfer site at surfaces $'$ and $''$, respectively. *K*s, mass law constants; *k*s, rate constants. According to this diagram, the anion species *a* and *b* compete for the various forms of the transfer site (protonated and deprotonated, facing compartment $'$ and $''$). There is no slippage (i.e., there are no transitions $r \rightleftharpoons s$ and $\rho \rightleftharpoons \sigma$). The transitions k_{15}, k_{51}, and k_{1122} k_{2211} are slow compared with the transitions k_{1155}, k_{5511} and k_{12}, k_{21} respectively. It has been suspected that after dansylation the transitions k_{15} and k_{51} may take place at a rate exceeding those of the transitions k_{1155} and k_{5511}. The reaction network is based on *Legrum* et al. (1980) and *Gunn* and *Milanick* (1982). The equations describing anion equilibrium exchange as a function of pH and anion concentration for a single divalent anion species in cells and medium have been published by *Legrum* et al. (1980)

ficulty that SO_4^{2-} equilibrium exchange passes through a maximum while Cl^- equilibrium exchange does not. This suggests that the rate of SO_4^{2-} equilibrium exchange is the result of two counteracting effects: an activation of sulfate transport by protonation of the substrate binding site and an inhibition by the protonation of another proton-binding site that is allosterically linked to the substrate-binding site. To differentiate between the effects of proton binding to these two sites, *Milanick* and *Gunn* (1984) measured the pH dependence of SO_4^{2-} net uptake by chloride-containing red blood cells at fixed intracellular pH in all sulfate media of varying pH. Under these conditions, the degree of protonation of the inward-facing proton-binding modifier sites is constant and all transfer sites are recruited to the outer membrane surface (see p. 106ff.). Hence changes of the protonation of the band 3 protein are confined to changes at an outward-directed proton-binding site. *Milanick* and *Gunn* observed that the pH maximum seen at equilibrium exchange is now replaced by a monotonic increase of transport with decreasing pH until a plateau is reached. The relationship between transport and external pH could be described by a single dissociation constant of p*K* 5.5 at 0°C and 5.9 at 22°C. The p*K* value of 5.5 is close to the p*K* that governs the inhibition of Cl flux as measured at 0°C. This provided strong support for the idea of an interconversion of the band 3 protein by protonation from a carrier for a monovalent anion into a carrier for a divalent anion.

Although H^+/SO_4^{2-} cotransport is the predominant mode of sulfate transport across the red cell membrane, SO_4^{2-} ions can also be transported without accompanying proton (*Milanick* and *Gunn* 1984). This possibility, first envisaged by *Legrum* et al. (1980), plays a role in the interpretation of the effects of dansylation of the red cell membrane on SO_4^{2-} transport.

After dansylation of the human red cell membrane the pH dependence of the band 3 protein-mediated (i.e., H_2DIDS-sensitive) SO_4^{2-} transport is changed. The maximum between pH 6 and 7 disappears and is replaced by a plateau above pH 7.0. Thus a pattern appears that is characteristic for monovalent anion transport. This suggests a loss of the ability of the band 3 protein to discriminate between monovalent and divalent anion species. Regardless of their charge, the anions are treated like monovalent anion species. The appearance of the plateau in the dansylated red cells implies that the enhancement of SO_4^{2-} equilibrium exchange increases with increasing pH. This is due to the decrease of the rate of H^+/SO_4^{2-} cotransport in the untreated control cells: the lower the rate of transport in the control, the higher the enhancement required to reach the plateau levels. At sufficiently high pH values, a several hundredfold enhancement above the untreated control has been observed (*Lepke* and *Passow* 1982; *Legrum* et al. 1980; *Berghout* et al. 1984; see p. 166). The reason for the loss of discrimination between mono- and divalent anions is unclear. The cotransport of H^+ and SO_4^{2-} continues to exist as in the untreated membrane over the whole pH range that is accessible for cotransport studies (pH 6.0–7.2). Perhaps, the changes seen in the dansylated red cells at higher pH values represent an additional transport of SO_4^{2-} without an accompanying proton, suggesting that such mode of operation of the transport protein is feasible in principle, although too small to be detectable under physiological conditions (*Berghout, Legrum,* and *Passow,* unpublished work, 1985). Technical problems have prevented us so far from establishing whether the enhanced SO_4^{2-} transport does or does not contribute to the conductance of the membrane.

The kinetics of phosphate transport are essentially similar to those of sulfate transport, including the pH dependence with its characteristic maximum. Similar to sulfate transport, V_{max} of phosphate transport is much more affected by the pH variations than $K_{1/2}$ (*Schnell* et al. 1981). The interpretation of the data is complicated by the fact that over the physiological pH range there are variations of the ratio between monovalent and divalent phosphate ions. Attempts have been made by *Runyon* and *Gunn* (1984) to determine the relative contributions of $H_2PO_4^-$ and HPO_4^{2-} to the total phosphate flux as measured by means of $^{32}PO_4$. In a preliminary report they reach the conclusion that the penetration rate of the divalent species is negligible compared with that of the monovalent species. The different effects of dansylation on monovalent and divalent anion exchange were used by *Berghout* et al. (1985) to address the same question. After dansylation at a fixed pH, the phosphate exchange was measured at a range of pH values. It was partially inhibited at low pH where $H_2PO_4^-$ predominates and considerably enhanced at high pH where HPO_4^{2-} prevails. The result of the two opposite effects is the nearly complete disappearance of the original pH dependence of phosphate equilibrium exchange which shows in untreated red cells a maximum around pH 6.3. Since dansylation is known to inhibit transport of monovalent Cl^- and to enhance transport of divalent SO_4^{2-}, the effect of dansylation on both monovalent and divalent phosphate ions suggests that both ion species are capable of penetrating.

In this context it should be mentioned that in the red cell of a fish, *Salmo irideus* (trout), divalent anion equilibrium exchange as measured by means of SO_4^{2-} in the pH range 6–8 is nearly independent of pH and that H^+/SO_4^{2-} cotransport can only be observed at temperatures far above the trout's physiological temperature range (i.e., above 40°C). The red cell membrane of the fish contains a protein that combines with H_2DIDS and migrates at the same location and with the same diffuse appearance as the band 3 protein of the mammalian red cells (*Romano* and *Passow* 1984; *Passow* et al. 1984a). Thus, it may be concluded that there exist homologues of band 3 in which the capacity to mediate H^+/SO_4 cotransport is less developed than in the red cells of other vertebrates and which behave somewhat similarly to the dansylated band 3 protein in the human red cell membrane. There exists, however, one difference. Even after dansylation, in the human red cell divalent anion transport still shows the same high activation enthalpy of 30 kcal mol^{-1} as in the untreated membrane (*Legrum* et al. 1980) whereas in the fish red cell the activation enthalpy for the band 3-mediated transport is about one-half the activation enthalpy determined in mammalian red cells (*Romano* and *Passow* 1984; see also *Obaid* et al. 1979). Interestingly enough, in the red cells of the urodele *Necturus* and the amphibian *Xenopus* H^+/SO_4^{2-} cotransport is easily observed (*Passow* et al. 1984a). In the latter, the activation enthalpy of SO_4^{2-} equilibrium exchange is similar to that in mammalian red cells at temperatures below 18°C and above that temperature more similar to that in the fish red cells (*Morgan* and *Passow,* unpublished work).

In conclusion of this section, it may be added that protons are not only cotransported with divalent anion species (*Jennings* 1978). In CO_2-free media, containing Cl^- or other halides as the principal anion species present, the proton equilibration after an acid or alkaline pulse can be described by the equation $J = k_H (H_i^+ Cl_i^- - H_o^+ Cl_o^-)$, where k_H is a rate coefficient that varies only little over the pH range 5.5–7.0. This equation can be derived on the assumption that the monovalent anion exchange system mediates an H^+/Cl^- cotransport. The rate of the net H^+ movements decreases in the sequence $Cl^- > Br^- > I^-$ parallel to the relative rates of Cl^-, Br^- and I^- self-exchange. It can be reduced to very low values by H_2DIDS and shows a temperature dependence that can be represented by an activation enthalpy of 27 kcal mol^{-1} between 5°C and 13°C and of 16 kcal between 13°C and 37°C. The absolute values of the rate constant for the net movements of H^+ and Cl^- are much lower, however, than the values for the rates of the corresponding Cl^- equilibrium exchange (*Jennings* 1978).

4 Molecular Basis of Kinetics

In the preceding section the kinetics of anion transport were discussed in terms of anion binding and translocation by a "transfer site" whose activities are subject to the influence of allosterically linked "modifier sites."

Studies from this and other laboratories have made it clear that the transfer site is composed of many amino acid residues that are contributed by different segments of the peptide chain. These residues constitute a region similar to the active center of an enzyme, which is capable of undergoing cyclic variations of its structural organization during substrate binding and translocation.

The modifier sites may consist of single amino acid residues that exert their influences upon the transfer site by changes of their state of protonation or oxidation or after covalent reaction with suitable modifiers. However, similar to substrate binding, the binding of noncovalently reacting modifiers may require the involvement of many different amino acid residues. In such cases, the binding may not simply depend on the spatial arrangement of the amino acid residues in the unmodified protein molecule, but also on the properties of the modifying agents. Similar to the substrates at the transfer site, the modifiers may interact with the amino acid residues at their respective binding sites by electron exchange or by ionic, dipolar, hydrophobic, or other forces. This may lead to a rearrangement of flexible regions of the peptide chain until a new configuration of minimal free enthalpy is obtained (induced fit). Thus the exploration of both binding and effects of specific modifying agents may provide information about the local properties of specific regions of the transport protein and about the relationships between these regions and the transfer site.

Among the amino acid residues that constitute the transfer site and the physiologically significant modifier sites (i.e., modifier sites that combine with substrate ions and protons), only a few have been identified so far. Since, in addition, the folding of the peptide chain of the transport-related domain of the band 3 protein is still largely unknown, our current views on the molecular basis of anion transport are essentially limited to inferences drawn from studies of the effects of chemical and enzymatic modification on the kinetics of the transport process. In these studies, the use of H_2DIDS and related stilbene disulfonates proved to be of singular value (*Knauf* and *Rothstein* 1971; *Cabantchik* and *Rothstein* 1974a). These nonpenetrating compounds inhibit anion transport at the outer membrane surface, but are ineffective at the inner membrane surface (*Passow* et al. 1975; *Zaki* et al. 1975; *Kaplan* et al. 1976; *Barzilay* and *Cabantchik* 1979b; and others). Their binding site seems to overlap with the outer entrance to the anion channel. It is very susceptible to controlled chemical and enzymatic modification and has been explored in great detail. Much of what follows is concerned, therefore, with results of these explorations and the conclusions drawn from them concerning the structure of the anion channel and the molecular mechanism of anion transport. Although some slowly penetrating compounds are known that inhibit anion transport by reacting at the inner membrane surface (e.g., APMB, NAP-taurine), their site and mode of action has been studied much less extensively than that of the stilbene disulfonates at the outer membrane surface. For this reason almost nothing is known about the properties of the channel inside the band 3 protein and of its internally facing orifice. So far, no more than a few unrelated observations have become available that shed some light on these other aspects of the mediation of anion transport by the band 3 protein.

4.1 Stilbene Disulfonate Binding Site

4.1.1 Relationship Between Binding Sites for Substrates and Disulfonates

4.1.1.1 Evidence from Measurements of Substrate Binding by Nuclear Magnetic Resonance Spectroscopy

Direct evidence for binding of stilbene disulfonates and substrates to overlapping sites comes from the study of anion binding by nuclear magnetic resonance (NMR) spectroscopy. The physiologically most important anion species, $^{35}Cl^-$, yields in aqueous solutions a single NMR resonance of Lorentzian shape. When Cl^- is bound to a protein, the shape of the resonance spectrum remains virtually unaltered but the line width at half height of the resonance line is increased. The line width observed is the result of the linear superimposition of the contributions of the line widths of free and bound chloride, whereby different species of chloride-binding sites may make different contributions at equal fractional occupancy. This well-known observation forms the basis for measurements of Cl^- binding to the band 3 protein both in leaky (*Shami* et al. 1977; *Falke* et al. 1984a) and resealed red cell ghosts (*Falke* et al. 1984b). The results obtained can be summarized as follows:

There exist two classes of chloride-binding sites, one with low, the other with high affinity. The chloride bound to the high-affinity sites is released upon addition of the stilbene disulfonate DNDS. From the dependence of chloride release on DNDS concentration, an apparent dissociation constant for DNDS binding could be calculated. A value of 6.4 μM was obtained, which is close to the value of 4.4 μM, derived from inhibition of chloride flux as measured at the same chloride concentration of 150 mmol liter^{-1}. I^-, Br^-, F^-, and HCO_3^- were found to compete with Cl^- for the high-affinity site. The sequence of the relative affinities for binding as determined by NMR was the same as derived from the competitive inhibition of Cl^- transport by the anion species mentioned. The findings described suggest that the binding site for DNDS overlaps with the anion transfer site (*Falke* et al. 1984a).

The chloride bound to the low-affinity binding site is not released by DNDS and the relative affinities of I^-, Br^-, Cl^-, F^-, and HCO_3^- to this site are different from those to the high-affinity site. This is in accord with the view that the low-affinity site binds substrate ions that are not transported by the protein molecule. It remains to be established whether or not the low-affinity binding site is identical to Dalmark's modifier site (*Falke* et al. 1984a).

4.1.1.2 Evidence from Inhibition Kinetics

Obtaining evidence for the overlap of the binding sites for substrates and stilbene disulfonates from kinetic studies was facilitated by the fact that the binding of H_2DIDS proceeds in two consecutive steps (*Cabantchik* and *Rothstein* 1974a; *Lepke* et al. 1976): The first is reversible and complete in less than 1 s. It leads to inhibition of transport. The second requires minutes or hours, depending on pH and temperature. It is irreversible and consists of covalent reactions of its two isothiocyanate groups with two lysine residues in two adjacent segments of the peptide chain, which leads to the formation of an intramolecular cross-link (*Jennings* and *Passow* 1979). The covalent reactions produce no further effect on transport.

At low temperature (0°C) and low pH (7.2) the rate of covalent bond formation of reversibly bound H_2DIDS can be reduced to the point where it is possible to demonstrate competition with Cl^- before the irreversible reaction interferes (*Shami* et al. 1978). In view of the extremely low apparent K_I of 0.046 mmol liter^{-1} it was necessary to use high concentrations of extracellular Cl^- to obtain measurable competition. Thus, although the data provided an important clue it was not possible to exclude that the observed effects were due to "allosteric competition" (*Passow* et al. 1980a,b), involving two distinct but allosterically linked sites rather than direct competition for the same site. Subsequent work has shown that tetrathionate, an anion with two sulfonate groups, joined together by a dithiol bridge showed a mixed type of inhibition which could best be explained by the involvement of an anion-binding modifier site (*Deuticke* et al. 1978). This site could, perhaps, be identical to the outward-facing NAP-taurine-binding site discussed on p. 164, which is different from Dalmark's modifier site but, like Dalmark's site, capable of substrate binding. However, in more recent work with noncovalently binding stilbene disulfonates (see Fig. 14) that were radioactively labeled (DNDS, *Fröhlich* 1982), fluorescent labeled (DBDS, *Rao* et al. 1979), or spin labeled (NDS-tempo, *Schnell* et al. 1983), it was possible to study competition at low substrate concentration where anion binding to the inhibitory modifier site plays little if any role. Both the inhibition of transport as well as the directly measured binding of the inhibitor showed clear evidence of competition for the transfer site[8] with indistinguishable values for the apparent K_I's for inhibition of anion exchange and the dissociation constants for stilbene binding.

---

8 See footnote to p. 134

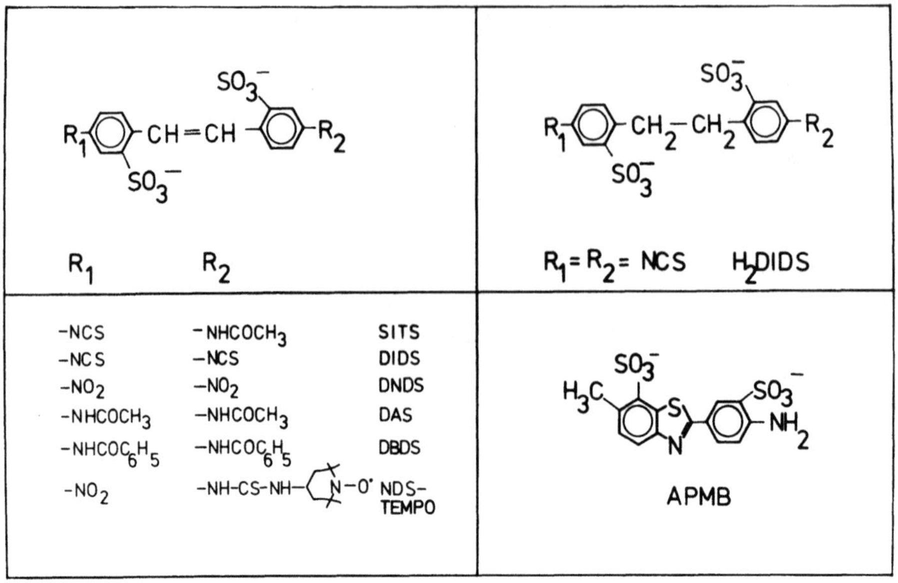

Fig. 14. Stilbene disulfonate derivatives and related compounds

Work of *Wieth* and *Bjerrum* (1983) suggests that the stilbene disulfonate binding site does not only overlap with the transfer site but also with the substrate-binding modifier site that is located in the outer cell surface and different from Dalmark's site. The Danish authors measured the inhibition by DNDS of chloride self exchange at 0°C at pH values up to about 12.0. Their measurements were performed at a chloride concentration in the external medium that was sufficiently high to saturate nearly the outward-facing transfer site, but much too low to saturate the modifier site. Under these conditions, in the controls without DNDS, the Cl^- transport remains unchanged up to about pH 11 and decreases when this pH is exceeded. There was no indication of the stimulation of Cl^- transport that is observed around pH 10.7 as a consequence of Cl^- release from the chloride-saturated modifier site when this site is deprotonated by increasing the pH at high Cl^- concentration in the medium (see p. 119). However, a stimulation of Cl^- transport was seen when the titration was performed in the same low Cl^- medium in the presence of DNDS at a concentration that produced about 50% inhibition at physiological pH values. According to the authors this suggests that the DNDS was released, at least in part, as a consequence of the deprotonation of the modifier site (p*K* 10.7) rather than of the transfer site and that the release from the modifier site leads to a decrease of inhibition of anion exchange.

Interestingly enough, at pH values exceeding those required to deprotonate the modifier site, the inhibition by DNDS at the transfer site also disappears, together with the gradual loss of the capacity to mediate Cl^- exchange due to the deprotonation of an arginine residue near the transfer site. Whatever the final explanation of this latter finding may be it is clear that the observations strongly suggest a close spatial proximity of transfer site and the outward-facing modifier site.

Table 3. Effects of varying the chloride gradient across the human red cell membrane on inhibitory potency of NAP-taurine and H_2DIDS. (*Knauf* et al. 1980)

	Cl_i^-/Cl_o^-	$K''_{1/2}$ (μM)
NAP-taurine	0.93	23.8
	5.33	10.7
H_2DIDS	0.99	0.12
	4.75	0.053

The external Cl^- concentration was 10 mmol liter^{-1} in all experiments. pH was 7.2. $K''_{1/2}$ indicates the inhibitor concentration in the external medium at which Cl^- exchange is inhibited by 50%

4.1.2 Recruitment of Binding Sites for Substrates and Stilbene Disulfonates

Knauf et al. (1980) (Table 3, lower half), *Furuya* et al. (1984), and *Fröhlich* (1982) investigated the stilbene disulfonate binding at a range of outward-directed Cl^- gradients across the membrane, thereby recruiting the transfer sites of the transport molecules into the outward-facing conformation. They obtained clear evidence for a recruitment of the stilbene disulfonate binding site: When the inner surface of the membrane is saturated with Cl^-, while the outer surface is unsaturated (i.e., $K_{101}/a' \ll K_{11}/a''$ in Eq. 15b, then $K''_{Iapp.} = K_{31}$ (see p. 108), i.e., equal to the true dissociation constant. Under this condition *Fröhlich* (1982) found for the inhibition of transport by DNDS and for DNDS binding virtually identical affinity constants of 80–90 nmol liter^{-1}. When, however, the measurements were done at equal chloride concentrations on both surfaces, K''_{Iapp} can be expected to differ from the true mass law constant for stilbene disulfonate binding, K_{303}. From Eq. 15b it follows for $a' = a'' = a$

$$K''_{Iapp} = \frac{K_{31}}{K_{11}} \cdot \frac{1+q}{q} \cdot (a + K_{1/2}) \quad \text{where} \quad K_{1/2} = \frac{K_{101} + qK_{11}}{1+q} \tag{21}$$

Fröhlich observed in fact that K''_{Iapp} was a linear function of the chloride concentration a in cells and medium, as predicted by Eq. 21. For $a = 0$, he obtained $K''_{Iapp} = 590$ nmol liter^{-1}, or about 6.5 times the true $K_I (= K_{31})$ of 90 nmol liter^{-1}. Assuming $K_{11} = K_{101}$ this corresponds to $K''_{Iapp} = K_{31} \cdot (1+q)/q$, where $q = k_{12}/k_{21} = 0.18$. From the slope of the straight-line relationship between K''_{Iapp} and a, he was able to calculate the affinity of chloride binding to the transfer site $K_{11} = K_{101} = 39 \pm 4$ mmol liter^{-1}. q and K_{11} are within the range calculated from the experiments of *Gunn* and *Fröhlich* (1979) and *Hautmann* and *Schnell* (1985), who found q =

0.05, K_{11} = 60 mmol liter^{-1} and q = 0.3–0.4, K_{11} = 15 mmol liter^{-1}, respectively (see p. 104/105).[9]

The value of K''_{Iapp} ($a' = a'' = a$) of 590 nmol liter^{-1} is similar to an earlier estimate of 870 nmol liter^{-1} by *Barzilay* and *Cabantchik* (1979a), who were the first to study the effects of DNDS on anion equilibrium exchange (see also the more recent work of *Eidelman* and *Cabantchik* 1983b).

The results discussed in this section indicate that the site for the binding of stilbene disulfonates can be recruited in parallel to the transfer site and that the conformer with outward-facing transfer site has a much higher affinity for the stilbene disulfonate than the conformer with inward-facing transfer site.

4.1.3 Stilbene Disulfonate Binding Site Only Partially Overlaps with the Substrate Binding Site

Although it is clear from the observations described above that the stilbene disulfonate binding site and the substrate site overlap, there is also evidence that stilbene disulfonate binding extends into regions of the band 3 protein that are not directly related to substrate binding. This follows from a comparison of the effects of papain on anion transport and DNDS binding (*Jennings* and *Adams* 1981): Papain treatment reduces chloride equilibrium exchange as measured at equal chloride concentrations inside and outside the red cells. Even after exhaustive treatment, the maximal degree of inhibition does not exceed 85%–90%. The residual flux is DIDS sensitive and hence mediated by band 3 protein. *Jennings* and *Adams* (1981) have analyzed these observations in terms of the Ping-Pong model of anion transport with the aim of identifying those parameters whose changes are essentially responsible for the effects. They first showed that the reduction of the flux at $[Cl^-]_i = [Cl^-]_o$ is the net result of a decrease of both $V_{max} = k_{12} \cdot \overline{RS}/(1+q)$ and $K_{1/2} = (K_{101} + qK_{11})/(1+q)$ (see p. 104). They then performed influx measurements under conditions where the transfer site is recruited outward. For this purpose, they placed the red cells in media in which most of the Cl^- was substituted by a nonpenetrating anion species. They maintained the intracellular Cl^- concentration at a level high enough to ensure that the influx is close to V_{max}. In contrast to the situation with equal Cl^- concentrations on both

9 As has been stated above, in Ping-Pong systems noncompetitive inhibition may be affected by variations of substrate concentration, except when $a' = a''$ and $K_{11} = K_{101}$ (see p. 109ff.). In view of the independent evidence for the equality of the affinity constants K_{11} and K_{101} provided by *Knauf* et al. (1981) and *Knauf* and *Mann* (1982), one may infer from the confirmation of Eq. 21 by *Fröhlich* (1983) (Fig. 8) that DNDS and the substrate actually compete for the same site

sides of the membrane, under this condition V_{max} is equal to $k_{12} \cdot \overline{RS}$ (see p. 106ff.). The authors now observed that the Cl^- equilibrium exchange in the papain-treated red cells was no longer inhibited as compared with a control that had been fluxed under the same conditions. If anything, the rate of uptake of radiochloride was somewhat enhanced. This indicates that anion binding to the transfer site is still feasible after papain treatment and that the inhibition observed at equal Cl^- concentrations in cells and medium is essentially due to an increase of q.

The unaltered survival of the substrate binding site after exposure to papain was further confirmed by measurements of the true affinity of the transfer site to the substrate SO_4^{2-}. The chloride-containing red cells were suspended in media containing a range of different SO_4^{2-} concentrations and the rates of net Cl^-/SO_4^{2-} exchange were determined. Using Eq. 13b the affinity constant (K_{11}) could be calculated and was found to be unaltered by the papain treatment.

The results obtained with substrates contrast with those obtained with stilbene disulfonates. After papain treatment, the residual Cl^-/Cl^- exchange can still be inhibited by stilbene disulfonates, but the concentrations required are much higher than in untreated cells. A quantitative estimate of DNDS binding showed a decrease of the true affinity (as determined from measuring the inhibition of net Cl^-/SO_4^{2-} exchange) by 1/12 after enzymatic treatment. Thus, the effects on substrate and stilbene binding are quite distinct although it is clear that both bind to overlapping sites.

It may be recalled that papain releases a 6-kDa piece of which about 1 kDa or less comes from the 17-kDa segment and about 5 kDa from the 35-kDa segment of the band 3 peptide. Evidently, this piece is important for stilbene disulfonate binding but not for substrate binding although it plays a role in the transition of the substrate-occupied transfer site from inward- to outward-oriented conformation.

4.1.4 Structural Features of the Stilbene Disulfonate Binding Site

A first glimpse at the structure of the H_2DIDS-binding site came from the study of the relationship between molecular structure of reversibly binding stilbene disulfonate derivatives and their inhibitory power. It was shown by *Barzilay* et al. (1979) that various substitutions at the two amino groups of the stilbene molecule altered the K_{Iapp} between 2 and 50 000 μmol liter^{-1}. These enormous variations could be attributed to the changes of electronic structure and hydrophobicity conveyed to the inhibitor molecule by the various substituents. Thus there existed a positive correlation with the Hammett constant σ, a measure of the capacity to exchange electrons ($r = 0.71$), and the Hansch constant π, a measure of hydrophobicity ($r = 0.94$). The two effects could be represented by a linear superimposition, giving rise to the relationship $-\log K_{\mathrm{Iapp}} = -0.73 + 1.01\,\sigma + 1.45\,\pi$ ($r = 0.998$).

The fluorescence enhancement of bound DBDS which competes with DNDS for the same binding site also indicates combination with the protein in a hydrophobic region, and the study of resonance energy transfer suggests a neighborhood of one or more tryptophane residues (*Rao* et al. 1979).

The hydrophobic stilbene disulfonate binding region seems to represent a cleft in the protein which extends some distance into the membrane. This is inferred from several different observations: (1) A spin-labeled derivative is immobilized after binding (*Schnell* et al. 1983). (2) Energy transfer measurements between the bound DBDS and fluorescent *N*-ethylmaleimide (NEM) derivatives covalently attached to a SH group on the 42-kDa segment of the band 3 protein yield a distance of only 34–42 Å, i.e., less than expected for the length of a segment of the peptide chain of a protein molecule that traverses the lipid bilayer (*Rao* et al. 1979). (3) Finally, H_2DIDS binding introduces negative charges into the cell surface but does not increase the net surface charge density of the red blood cell membrane as measured by cell electrophoresis (*Fuhrmann* and *Passow,* unpublished work). Even after removal of neuraminic acid by either neuraminidase or trypsin and the concomitant reduction of surface charge by 80% or more, and even in media of low ionic strength (about 1% of the normal value), when contributions to the surface charge of layers below the plane of shear should become easily apparent, none of the changes of zeta potential can be seen that one could expect to occur as a consequence of the introduction into the cell surface of two sulfonate groups per band 3 molecule.

The incorporation of the bulky stilbene disulfonates into the hydrophobic cleft near the outer cell surface leads to profound changes of the structure of the band 3 molecule as a whole. The changes can be monitored at the exofacial carbohydrate moiety, at the intrafacial 42-kDa domain, and in the surrounding lipids. Thus, H_2DIDS binding substantially reduces the adenosine-enhanced concanavalin A agglutination of the red cells (*Singer* and *Morrison* 1980), increases hemoglobin binding to the 42-kDa domain (*Salhany* et al. 1980), and modifies the calorimetric behavior of the lipid bilayer (*Appell* and *Low* 1982). These observations suggest that the H_2DIDS binding is associated with an induced fit which causes a major reorganization of the glycoprotein and its environment.

This conclusion is supported by *Fröhlich's* (1982) measurements of the temperature dependence of the mass law constant K_D for DNDS binding to the outward-recruited binding site. From the well-known thermodynamic relationship

$$RT\,ln\,K_D = \Delta H - \Delta S\,/\,T$$

he calculated $\Delta H = -0.5$ kcal ml^{-1} and $\Delta S = +30$ cal mol^{-1}. The data agree reasonably well with estimates by *Verkman* et al. (1981) for DBDS ($\Delta H = +1.4$ kcal mol^{-1}, $\Delta S = +26$ cal). The surprisingly large enthropy change associated with DNDS binding is possibly related to a change of hydration of DNDS or the anion-binding site, but it is quite likely that the large conformational changes of the transport protein described above are an important contributing factor (*Fröhlich* 1982).

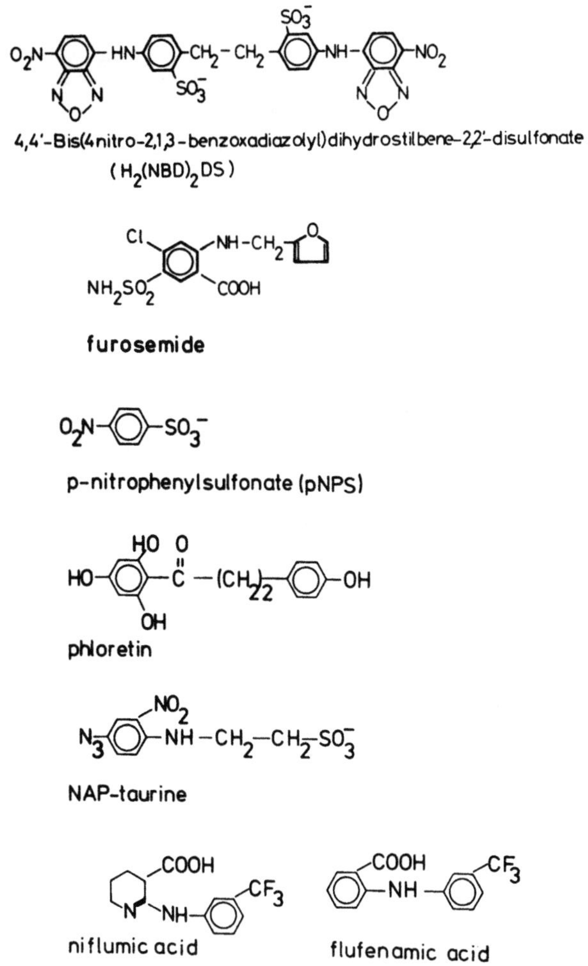

Fig. 15. Reversibly binding inhibitors of anion transport whose binding overlaps with that of the stilbene disulfonates

4.1.5 Interrelationship Between the Binding of Stilbene Disulfonate and Other Inhibitors of Anion Transport

There exist a number of inhibitors whose binding to the band 3 protein is mutually exclusive with the binding of the stilbene disulfonates (Fig. 15).

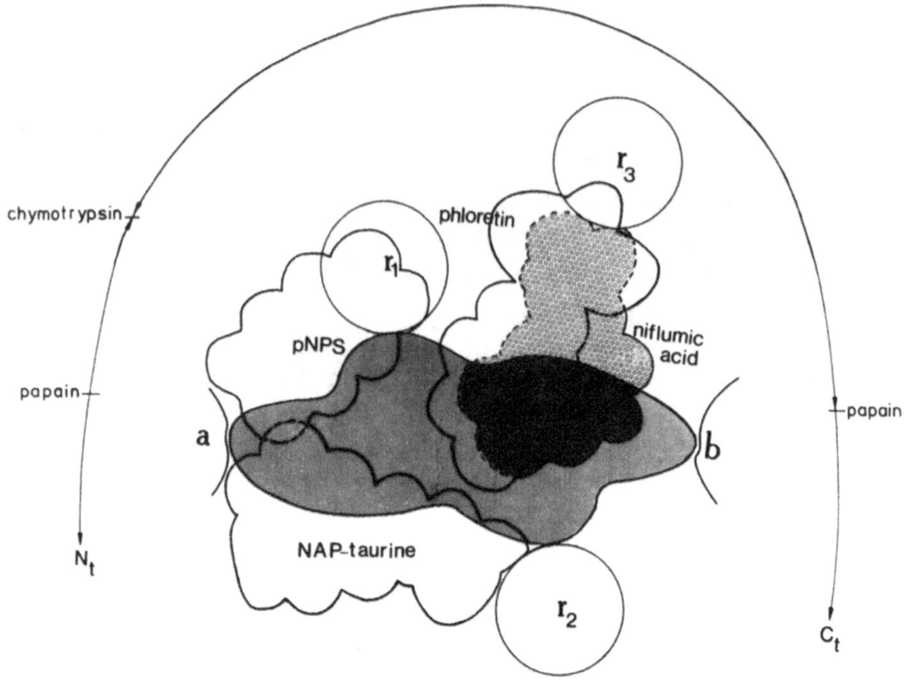

Fig. 16. Topological relationships between the binding sites of five different classes of inhibitors of anion transport at the stilbene disulfonate binding site: a hypothetical picture. *Dashed lines* indicate that information about overlap is missing. *a, b,* lysine residues in 17-kDa and 35-kDa segments, respectively. Distance from lys *a* to lys *b* is about 20 Å. r_1, r_2, r_3, presumably arginine residues. r_1 and r_2 are possibly constituents of transfer and modifier site, respectively. The *dark area* between *a* and *b* represents an H_2DIDS molecule. Cleavage by papain of the peptide chain at two locations that are about 5 kDa apart from one another reduces drastically the affinity for binding of DNDS (which would reside at the same location as H_2DIDS) and NAP-taurine but not the affinity for substrate binding. The papain-digestible area contains one or two carboxyl groups that modify anion transport. The picture is based on work from many laboratories, cited in the text

It is quite likely, therefore, that their site of action is located in the same hydrophobic region that accommodates the stilbene disulfonates. Their effects on anion transport as well as their mutual interactions have been studied in some detail. This provides some information on the structural organization of the stilbene disulfonate binding site and allows some provisional mapping of the locations of the sites of action of the various inhibitors relative to each other (Fig. 16). However, allosteric rather than steric interactions cannot be excluded entirely. In the case of allosteric interactions, the discussion presented below would provide a map of functional rather than topological relationships.

The published data about the noncovalent chemical reactions at the stilbene disulfonate binding site will be presented within the context in which the respective authors collected them. However, they will be discussed in relation to a hypothetical model of the structure of the stilbene disulfonate binding site (*Passow* et al. 1980a,b). This serves to pull together a considerable number of otherwise seemingly unrelated observations but does not imply that this model is believed to represent the final truth. The model assumes a cluster of four to five positively charged groups, two lysine, and two to three arginine residues that reside near the outer surface in a hydrophobic pocket of the band 3 molecule of a diameter of about 20 Å, the size of an H_2DIDS molecule. The data will be analyzed in an attempt to determine the locations of the individual residues and their specific functions in the transport process (see Fig. 16).

Phenylmonosulfonates are roughly equivalent to one-half of a stilbene disulfonate molecule. They are accepted as substrates by the transport system and thus are capable of combining with the substrate-binding site. This agrees with the finding that they inhibit inorganic anion transport competitively (*Barzilay* and *Cabantchik* 1979b). The apparent K_I's are, however, much higher than those of most stilbene disulfonates, suggesting that they are less firmly bound due to the absence of the additional negative charge. However, the lower charge alone cannot account for the lower K_{Iapp}. This follows from the observation that phenyldisulfonates do not bind much stronger than the phenylmonosulfonates. Apparently, the spacing of the two sulfonate groups in the stilbene molecule is important for conveying the extraordinarily high affinity to the substrate-binding site. The electrostatic binding of the phenylmonosulfonates is modulated by the same factors that play a role in stilbene disulfonate binding, i.e., hydrophobicity and the capacity to exchange electrons with the environment. Analogously to stilbene disulfonates, the inhibitory potency of a wide range of derivatives of phenylsulfonates was found to be related to the superimposed effects of hydrophobicity and electronic structure and could be represented by the expression $-\log K_{Iapp} = 3.97 + 1.04\,\sigma + 0.15\,\pi$, where σ and π designate, respectively, the Hammett and the Hansch constant. One recognizes that the influence of the electronic structure is about the same as with the stilbenes, while the influence of the hydrophobicity is much smaller (*Barzilay* et al. 1979; cf. p. 135f.).

The covalently binding isothiocyano phenyl sulfonate (pIPS) and the diisothiocyano stilbene disulfonate H_2DIDS have about the same electronic structure and lipophilicity as their respective noncovalently binding analogues in which the isothiocyanate groups are replaced by nitro groups. Thus the covalently binding isothiocyanates should react in the close vicinity of the noncovalently binding derivatives. H_2DIDS and pIPS establish covalent bonds preferentially with the lysine residue *a* on the 17-kDa fragment of band 3. It may be appropriate, therefore, to assume that the phenylsulfonates occupy that part of the H_2DIDS-binding site that is close to a positive charge near lys *a* (r_1 in Fig. 16). Perhaps, this charged group could be involved in substrate binding. Since the phenylsulfonates are competitive inhibitors that are themselves transported, the charged group could be identical with the arginine residue that is titrated at pH 12 (*Wieth* and *Bjerrum* 1982) and located in the 35-kDa region of the band 3 protein (*Bjerrum* 1983; *Wieth* et al. 1982c) (see p. 158ff.).

Niflumic acid (*Cousin* and *Motais* 1979) and a related compound, flufenamate (an anthranilic acid derivative) (*Cousin* and *Motais* 1982a,b), produce strong noncompetitive inhibition and hence do not seem to interfere directly with the substrate-binding site. Nevertheless, their binding is mutually exclusive with that of the stilbene disulfonates, indicating an allosteric relationship or an overlap of the respective binding sites. It has been shown that the carboxyl group of the inhibitors needs to be ionized in order to permit the binding. Again, both hydrophobicity and the capacity to accept electrons are the essential modulating factors for the relative inhibitory power of a wide range of derivatives (*Cousin* and *Motais* 1982b).

It is interesting to note that furosemide is also an anthranilate derivative that inhibits anion transport. The type of inhibition is mixed and a detailed analysis by *Brazy* and *Gunn* (1976) has led to the result that the negatively charged form of the drug reacts with two different sites. One of them seems to be an outward-facing site at which the substrate causes self-inhibition and which is not identical to Dalmark's modifier site. At the second site, furosemide noncompetitively blocks chloride transport and alters the apparent affinity for the transport site. *Cousin* and *Motais* (1982b) have demonstrated that this compound fits into the series of other anthranilate derivatives. Its inhibitory potency depends on the negative charge and can be predicted quantitatively when the hydrophobicity and the capacity for the formation of the charge transfer complexes are taken into account.

Since the anthranilate derivatives need to combine with a positive charge, but are noncompetitive inhibitors, it is most likely that they do not combine with the arginine residue involved in substrate binding (r_1). They also do not compete with inorganic anions for the self-inhibitory modifier site. Hence, the arginyl residue (r_2) that can be titrated at pH 11 and that forms an essential part of the modifier site (*Wieth* and *Bjerrum* 1982) also does not participate in niflumic acid binding. Perhaps it is appropriate to assign to niflumic acid and the anthranilate derivatives a place that overlaps with the site near lys *b* where the benzene rings of stilbene disulfonates or phenylsulfonates reside and to assume that the positive charge required to neutralize the negative charge of the noncompetitive inhibitors is located near to lys *b* (r_2, Fig. 16).

Phloretin's binding to band 3 is mutually exclusive with that of stilbene disulfonates (*Formann* et al. 1981; *Fröhlich* and *Gunn* 1982). Its inhibitory power, like that of a whole group of other acetophenone derivatives, does not significantly depend on its electronic structure, but primarily on lipophilicity and dipole moment (*Motais* and *Cousin* 1978). Thus it seems clear that its binding site is different from that of the phenylsulfonates and anthranilic acid derivatives, although a partial overlap cannot be excluded.

NAP-taurine binding is also mutually exclusive with the binding of stilbene disulfonates [reviewed by *Knauf* (1979) and *Macara* and *Cantley* (1981b)]. It combines covalently with the 17-kDa fragment, its site of noncovalent binding should be closer to Lys *a* than to Lys *b* and the point of covalent attachment after illumination could be, perhaps, Lys *a* or an amino acid residue in its vicinity. The reaction with the modifier site excludes binding to the arginine residue near Lys *a* that participates in phenylsulfonate binding and which is presumably part of the transfer site (r_1). Perhaps the binding site for NAP-taurine includes an additional arginine residue, which may be identical to the residue r_2 that has been titrated at pH 11 by *Wieth* and *Bjerrum* (1982) and which seems to be involved in the binding of the second sulfonic acid residue of DNDS (*Wieth* and *Bjerrum* 1983).

The binding of both NAP-taurine and niflumic acid is facilitated when a large outward-directed Cl^- gradient across the membrane is established (*Knauf* et al. 1980, 1984). Thus similar to stilbene disulfonate binding, the binding sites of these two inhibitors become more accessible when the transfer site is recruited into the outward-facing form (Table 3).

Finally, and again similar to stilbene disulfonate binding, treatment with papain reduces the affinity for niflumic acid binding to negligibly small values (*Cousin* and *Motais* 1982a).

In contrast to NAP-taurine, niflumic acid binding is strictly noncompetitive. This supports the contention (see Fig. 16) that the negative charge of niflumic acid that is required for binding is neutralized by a positive charge in the membrane that is not identical with the arginyl residue r_2, which seems to be involved in NAP-taurine binding. Thus, the characteristics of niflumic acid binding and effect suggest the existence of a third arginine residue (r_3) in the stilbene disulfonate binding region of the band 3 protein in addition to two other positively charged groups, one of which is part of the transfer site (r_1) and the other of which is part of a substrate-binding modifier site (r_2).

It is most likely that the three positively charged groups whose existence has been inferred from the work reviewed above are located in the same area as the lysine residues *a* and *b* that are involved in the covalent bond formation with the two isothiocyanate groups of H_2DIDS. However, these residues play no major role in noncovalent binding since, over the pH range 7–9, stilbene disulfonate binding is pH independent although the degree of protonation of Lys *a* and *b* can be expected to change (see below). Noncovalent stilbene disulfonate binding is reduced only when the arginine residues at the transfer and the modifier site become deprotonated, i.e., above pH 9 (vide infra) (*Wieth* and *Bjerrum* 1983).

4.1.6 Organization of the Stilbene Disulfonate Binding Site as Inferred from Study of Noncovalently Binding Inhibitors. A Hypothesis

Although the information about the mutual interrelationships among the binding sites of the various classes of inhibitors is still fragmentary, a schematic figure has been drawn which tentatively tries to summarize the results of the studies discussed above (Fig. 16).

Figure 16 shows that many data derived from such different sources as structure-activity relationships, inhibition kinetics, determinations of reversible binding, affinity labeling, and acid base titrations can be accommodated within a reasonably simple picture: a hydrophobic surface in a pocket at the outer surface of the transport protein with a diameter of about 20 Å (i.e., the length of a stilbene disulfonate molecule like DNDS or H_2DIDS), containing three positively charged groups in addition to two lysine residues involved in the intramolecular cross-linking. The positive charge within the cluster would not only produce a high local surface potential that attracts the substrate to the mouth of the channel. In addition, each of the charged groups would play some specific role, depending on its location and the hydrophobicity and electronic structure of the other amino acid residues in its vicinity. The positively charged residues r_1 and r_2 could be responsible for substrate binding to the transfer site and outward-facing modifier site, respectively, while the residue r_3 could be involved in the binding of anionic inhibitors that produce a noncompetitive inhibition. If the assignment of these roles were correct, one could further stipulate, on the basis of the titrations of *Wieth* and his associates, that at least r_1 and r_2 are arginine residues.

4.2 Kinetics of Stilbene Disulfonate Binding

4.2.1 Noncovalent Binding

Noncovalent binding was studied by means of the nonpenetrating, noncovalently binding probe 4,4′-dibenzoamide stilbene-2,2′-disulfonate (DBDS, see Fig. 14) whose fluorescence is enhanced by binding to hydrophobic surfaces (*Dix* et al. 1979). Both stopped flow and temperature jump techniques were applied to follow the time course of the reaction between the inhibitor and the band 3 protein. There was a rapid binding with low affinity and a subsequent slow process (with a rate constant of about 4 s^{-1}) which was attributed to a conformational change that locks the DBDS molecule in place (*Verkman* et al. 1983). The diphasic process was measured at a range of DBDS concentrations and hence varying degrees of occupancy of the stilbene disulfonate binding site. The results were interpreted on the assumption that the reaction with the protein is a two-step process: After rapid reaction of one DBDS molecule a slower conformational change takes place which then permits the reaction of a second DBDS molecule. Since each band 3 protein monomer only combines with one DBDS molecule, this involves the hypothesis that interactions take place between different monomers of the band 3 dimers or tetramers (*Verkman* et al. 1983). Currently existing evidence suggests that the monomers are independently operating units, which would argue against this hypothesis (see p. 169); but a final judgment would need further experimental data.

Perhaps, the kinetics described can be related to the existence of two different, interconvertible conformations of the H_2DIDS-binding site with different affinities for the inhibitor (see p. 167). The temperature jump would disturb the originally existing equilibrium between each of these conformers and the DBDS in the medium. This should induce a redistribution process which would involve besides rapid changes of DBDS binding a slow interconversion of the two conformeric states and thus give rise to a diphasic process. There exists a whole series of abstracts on the kinetics of stilbene disulfonate binding from Solomon's laboratory which should be consulted for more recent information (*Forman* et al. 1981, 1982; *Dix* et al. 1981; *Dix* and *Verkman* 1982; *Verkman* et al. 1981, 1982).

4.2.2 Covalent Binding

Covalent binding of the noncovalently bound H_2DIDS molecule takes place at two lysine residues called Lys *a* and Lys *b* which reside, respectively, in the 60-kDa and the 35-kDa chymotryptic fragments described above. It can be followed with $^3H_2$DIDS after cleavage of the band 3 protein with external chymotrypsin, which, as may be recalled, does not affect the transport function.

After reversible binding the H_2DIDS molecule (denoted *s*) resides somewhere between Lys *a* and Lys *b*, a situation which may be denoted by *asb*. The molecule has the option to react covalently either first with Lys *a* or with Lys *b*, which may be designated as *asb* → *a-sb* and *asb* → *a s-b*, respectively (the hyphens indicate covalent bonds). Those H_2DIDS molecules that first reacted with Lys *a* will further react with Lys *b* (*a-sb* → *a-s-b*) and those that reacted first with Lys *b* will further react with Lys *a* (*a s-b* → *a-s-b*). The cross-linked form *a-s-b* would increase at the expense of *a-sb* and *as-b*. These latter two forms will pass through maxima, while the cross-linked form will increase monotonically, until all fragments exist in the form *a-s-b* (Fig. 17). The predictions of this "basic model" can be formulated mathematically and verified experimentally

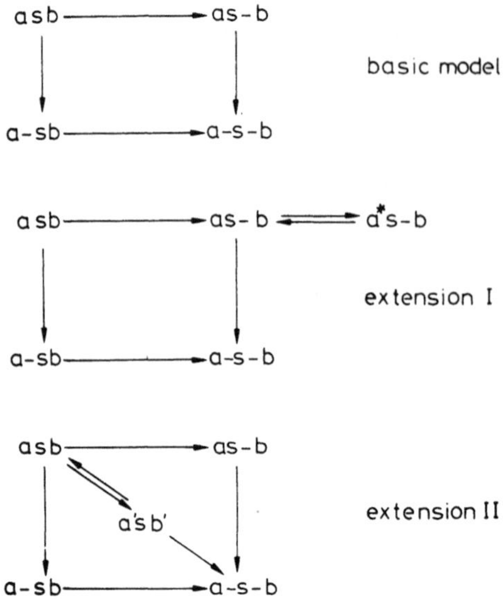

Fig. 17. Reaction diagram for a cross-linking reaction. *a* and *b* represent lysine residues on the 17-kDa and 35-kDa segment, respectively, of the band 3 protein. The *hyphens* indicate covalent bond formation between *s* and either *a* (*a-sb*) or *b* (*as-b*) or the formation of a cross-link (*a-s-b*). *a*s-b* and *a′sb′* are products that are incapable of covalent bond formation. For discussion see text

(*Passow* et al. 1982). H_2DIDS binding to the 35-kDa and 60-kDa fragments passes in fact through maxima and the formation of the cross-linked 95-kDa product continues monotonically (Fig. 18).

When one tries, however, to fit the data to the equation that describes the expected behavior quantitatively, deviations become apparent (Fig. 18a).

Among a number of extensions of the basic model that could account for these deviations two were found that can be fitted reasonably well to the data (Fig. 17) (*Kampmann* et al. 1982). The first postulates that in addition to the two reaction pathways (1) *asb* → *a-sb* → *a-s-b* and (2) *asb* → *as-b* → *a-s-b* there exists a third pathway where a simultaneous reaction between Lys *a*, Lys *b*, and the reversibly bound *s* takes place: (3) *asb* → *a-s-b*. For statistical reasons a simultaneous reaction of *s* with two independently oscillating amino groups *a* and *b* is unlikely. This extension involves, therefore, the tacit assumption that there exists a specific configuration of the H_2DIDS binding site (*a′sb′*) where a nearly simultaneous reaction with *a* and *b* becomes feasible: *asb* ⇌ *a′sb′* → *a-s-b*. The other extension of the model stipulates that the lysine residue *a* may exist in two states: one in which it is reactive (*a*) and another in which it is nonreactive (*a**) and that a slow transition between the two states affects the rate of the reaction:

$$\begin{array}{ccccc} asb & \rightarrow & as\text{-}b & \rightarrow & a\text{-}s\text{-}b \\ \downarrow\uparrow & & \downarrow\uparrow & & \\ a^{*}sb & \rightarrow & a^{*}s\text{-}b & & \end{array}$$

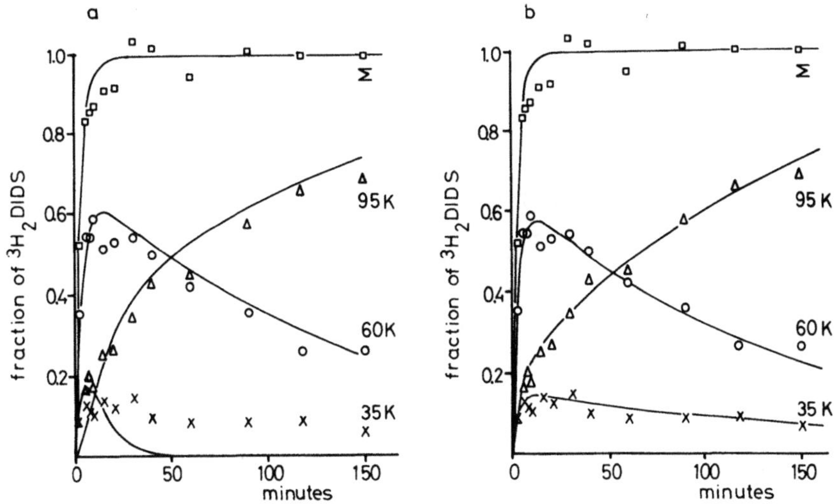

Fig. 18a,b. Time course of cross-linking with H_2DIDS of the chymotryptic 60-kDa and 35-kDa fragments of band 3. **a** Nonlinear least squares fit of the basic model (see Fig. 17) to the data points. **b** Fit of the extension I of the model assuming the existence of the nonreactive form ***a*s-b***. Chymotrypsinized red cells in 20 mmol liter^{-1} ethylenediaminetetraacetate (EDTA), 130 mmol liter^{-1} NaCl, pH 8.0, temperature 37°C, H_2DIDS concentration 20 μmol liter^{-1}. The uppermost curve (Σ) in both **a** and **b** represents the sum of all forms. *Ordinate*, fraction of total H_2DIDS bound per cell that is associated with the fragments indicated in the figure. *Abscissa*, time in minutes

It should be noted that although formally somewhat different from the preceding extension it also involves the assumption that the H_2DIDS binding site may exist in two different conformeric states: *asb* and *a*sb*.

Evaluation of the data in terms of both extensions yielded good fits, with nearly identical sums for the least squares of the errors. Nevertheless, we prefer the latter extension since it is supported by independent observations about the capacity of Lys *a* to exist in an exposed and buried state (Fig. 18b).

Within the scheme of the cross-linking reaction, it is experimentally feasible to isolate the partial reaction: *a-sb* → *a-s-b*. This is due to the fact that at low pH H_2DIDS reacts much faster with Lys *a* than with Lys *b*. Therefore, after exposure of the red cells to H_2DIDS at low pH for a short time, most of the inhibitor is covalently bound in the form *a-sb*. After removal of the H_2DIDS in the medium and of the reversibly bound H_2DIDS by washing with bovine serum albumin, most of the H_2DIDS that remains attached to the membrane exists in the form *a-sb*. The transition of this form into *a-s-b* can now be followed at any desired pH without the superimposition of the other reactions described above.

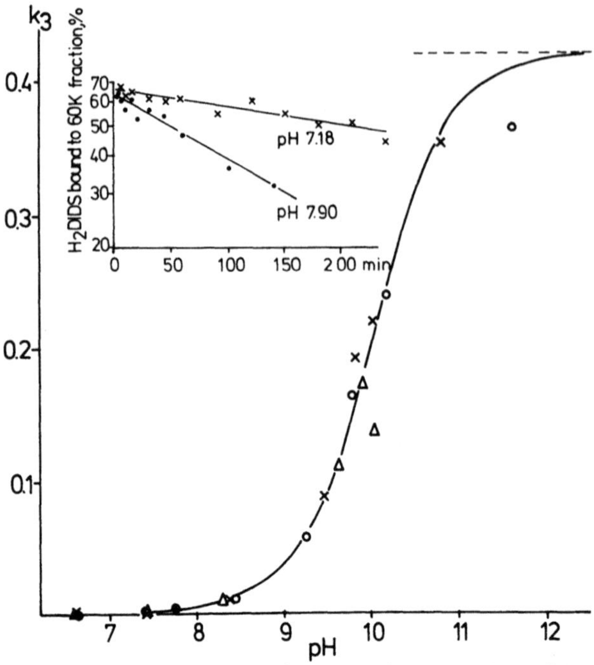

Fig. 19. pH dependence of the covalent reaction of Lys *b* with one of the isothiocyanate groups of H_2DIDS. *Ordinate*, first-order rate constants of the transition *a-s b* → *a-s-b*. *Abscissa*, pH (37°C). *Insert*, semilog plot of the time course of disappearance of *a-s b* as a consequence of the formation of *a-s-b*

The isolated reaction *a-sb* → *a-s-b* is of the first order (Fig. 19). The observed rate constants agree quite well with the corresponding rate constants derived from the computer evaluation of the complex time course of the whole reaction sequence of which the reaction *a-sb* → *a-s-b* forms one of the steps involved. The agreement was not only satisfactory when the rate constants were calculated by means of one or other of the two extended versions of the model, but even when the simple basic model was used that does not suffice to give a good fit for the full time course of the cross-linking reaction. This confirms that the reaction sequence *asb* → *a-sb* → *a-s-b* (which is common to the basic model and its extensions) is correctly described, while the description of the other branch of the reaction sequence (*asb* → *as-b* → *a-s-b*) represents an oversimplification. What follows below is exclusively confined to the isolated reaction *a-s b* → *a-s-b* and deals with the pH dependence of its first-order rate constant.

The isothiocyanate groups of H_2DIDS can react only with the deprotonated form of the amino group of a lysine residue $-NH_2 + SCN-R \rightarrow -NH-CS-NH-R$, but not with the protonated form NH_3^+. The reaction rate should, therefore, vary parallel to changes of the dissociation equilibrium $NH_3^+ \rightleftharpoons NH_2 + H^+$. Thus, the pH dependence of the rate constants of the

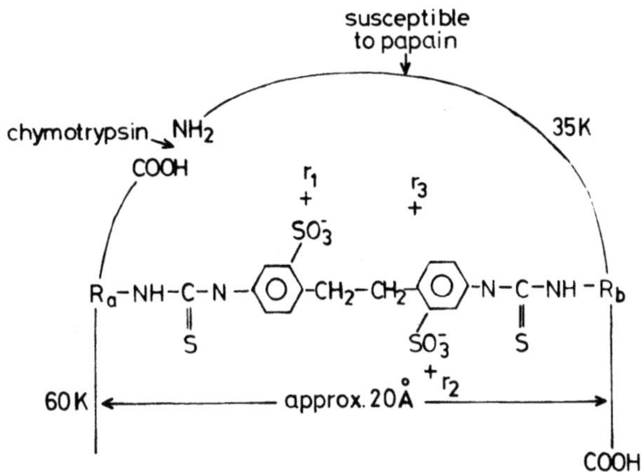

Fig. 20. The H_2DIDS binding site, as inferred from measurements of the time course of covalent H_2DIDS binding to Lys *a* and Lys *b*. (*Passow* et al. 1980a,b)

reaction *a-sb* → *a-s-b* should be governed by the p*K* value of the dissociation equilibrium for Lys *b*.

The rate constants measured at the various pH values can be fitted by a nonlinear least squares method to an S-shaped curve that pertains to a single p*K* value of 10.0 (Fig. 19). This is close to the expected value for the ϵ amino group of a lysine residue at high dielectric constant in an electrically neutral environment (p*K* 10.5, *Tanford* 1962). Unfortunately, the p*K* value of Lys *a* cannot be determined with the technique described. Only a rough estimate is possible, which suggests a p*K* value between 7 and 8 (*Kampmann* et al. 1982).

It is useful to discuss the implications of these results with respect to the structure of the H_2DIDS binding site using Fig. 20. This figure shows an H_2DIDS molecule that has established thiourea bonds between Lys *a* and Lys *b*, thereby bridging a distance of about 20 Å.

We shall first consider the finding that the p*K* of Lys *b* was close to 10.0. Although this value agrees with the p*K* value of a lysine residue in an electrically neutral environment, we are not entitled to believe that the dissociating amino group is indeed located in such an environment. The reason is the following: The H_2DIDS molecule possesses two SO_3^- groups that are introduced into the immediate vicinity of Lys *b*. They should increase the p*K* value of Lys *b* far above the observed value of 10. Since this is not the case, we may suspect that the two negative charges of H_2DIDS are compensated for by at least two positive charges near Lys *b*. Recent work in our laboratory (*Zaki* 1981, 1982, 1983, 1984; *Zaki* and *Julien* 1983, 1985) and in Copenhagen (*Wieth* et al. 1982a,c; *Bjerrum* et al. 1983; *Wieth* and *Bjerrum* 1982) suggests that arginine residues are involved in anion binding and transport by band 3 protein. The charged groups that we postulate to exist near Lys *a* and Lys *b* may be identical with these arginine residues (see below). Thus, within a distance similar to the size of an H_2DIDS molecule there seems to be located a cluster of at least four, perhaps even five, charged groups, two lysine residues, and two or three additional residues that are possibly arginine residues (*Passow* et al. 1980a,b). At least the lysine residues reside on different segments of the peptide chain, but in close juxtaposition (*Jennings* and *Passow* 1979). It is reassuring that the inferences drawn from the studies of reversible stilbene disulfonate binding discussed in a preceding section can be accommodated within this general picture (see Figs. 16 and 20).

4.3 Chemical Modification of Specific Amino Acid Residues

4.3.1 Lysine Residues

Information about the functional significance of one or both of the two lysine residues that are involved in the thiourea bond formation with the two isothiocyanate groups of H_2DIDS can be derived from work involving amino groups modifying procedures: reductive methylation (*Jennings* 1982a; *Jennings* and *Nicknish* 1984), dinitrophenylation (*Passow* et al. 1980a,b; *Rudloff* et al. 1983) and, possibly, treatment with pyridoxal-5-phosphate (*Nanri* et al. 1983).

4.3.1.1 Reductive Methylation

Repeated treatments with formaldehyde and $NaBH_4$ lead to a modification of both Lys *a* and Lys *b*, although the reaction with the latter is more complete than with the former. Reversible binding of H_2DIDS still takes place, but the covalent reaction with Lys *a* is partially, and with Lys *b* completely, prevented. Chloride equilibrium exchange is inhibited by 70%–80%. This could be attributed to the modification of Lys *b*. When the transport protein is recruited into the outward-facing conformation by suspension of the Cl^--containing cells in a medium containing the more slowly penetrating Br^-, the affinity of Br^- to this conformation can be measured (Eq. 13). It was found to be unchanged in spite of the unaltered inhibition of about 80%. This suggests (see p. 107f.) that the methylation affects the rate of the conformational changes associated with anion translocation but not anion binding (*Jennings* 1982a).

4.3.1.2 Dinitrophenylation

Dinitrophenylation of the red cell membrane produces inhibition of anion equilibrium exchange as measured with $^{36}Cl^-$ or $^{35}SO_4^{2-}$. The band 3 protein carries many amino acid residues that are susceptible to modifica-

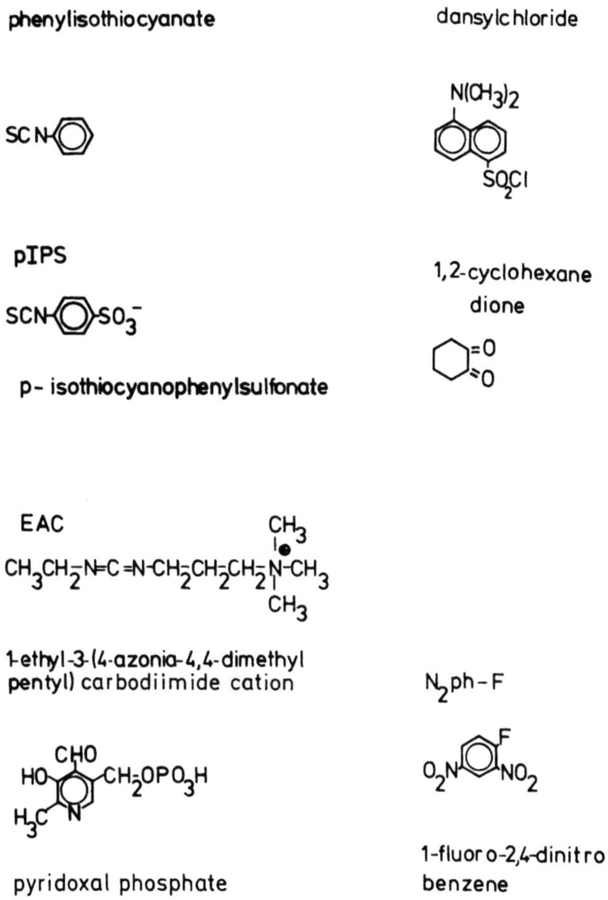

Fig. 21. Covalently binding chemical modifiers of anion transport

tion by 1-fluoro-2,4-dinitrobenzene (N_2 ph-F, see Fig. 21), including 27–28 lysine residues. Nevertheless, about 80% inhibition is achieved when no more than 2 dinitrophenyl residues are incorporated per band 3 molecule. At 95% inhibition, the stoichiometrical ratio is about 6:1. Thus, some amino acid residues are much more reactive than others (*Rudloff* et al. 1983).

When the red cells are treated with N_2 ph-F in the presence of a stilbene disulfonate, then the binding of one N_2 ph residue per band 3 molecule is prevented (*Passow* et al. 1975; *Zaki* et al. 1975). This indicates that one of the N_2 ph-F binding amino acid residues is part of the stilbene disulfonate binding site or allosterically linked to it. The N_2 ph-F binding site is located in the 17-kDa segment of the band 3 protein and could be identified by amino acid analysis as a lysine residue (*Rudloff* et al. 1983). It is presumably identical to Lys *a*, to which the isothiocyanate derivatives of stilbene

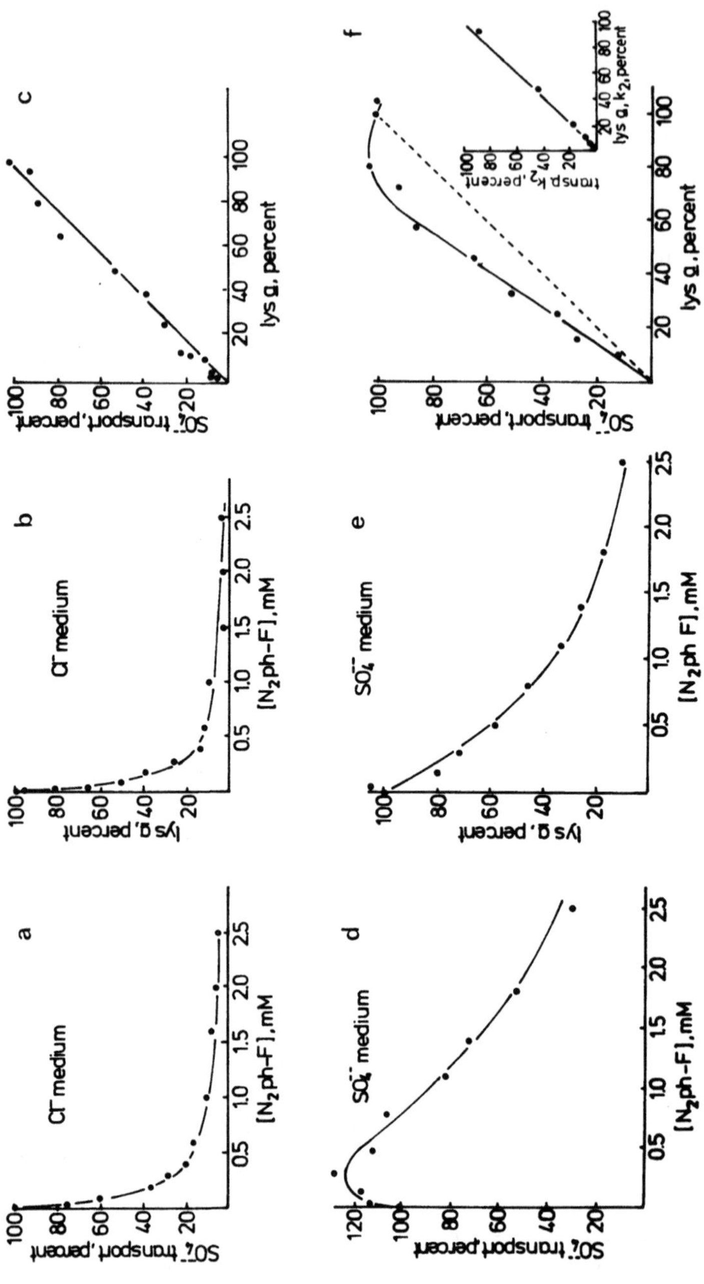

Fig. 22a–f. Effect of exposure to 1-fluoro-2,4-dinitrobenzene (N_2ph-F) for 30 min at 37°C on subsequently measured sulfate transport and the capacity of Lys *a* in band 3 to bind $^3H_2$DIDS. The latter is a quantitative measure of the residual number of Lys *a* residues that had not been modified by dinitrophenylation. Dinitrophenylation of ghosts in the presence of 130 mmol liter^{-1} NaCl, 20 mmol liter^{-1} EDTA, 1 m*M* liter^{-1} Na_2SO_4, pH 7.0 **a–c**; or 108 mmol liter^{-1} Na_2SO_4, 20 mmol liter^{-1} EDTA, pH 7.0 **d–f**. Equal composition of the media inside and outside red cell ghosts. **a, d** Sulfate flux as a percentage of flux in untreated ghosts. **b, e** H_2DIDS binding to band 3, per cell, as a percentage of H_2DIDS binding to untreated ghosts. In **a, b, d**, and **e**, the drawn lines are nonlinear least squares fits of the sums of two exponentials to the data points. **c, f** Flux from **a** and **d** plotted against corresponding H_2DIDS binding capacity from **b** and **e**, respectively. Transport equals $A_1 \exp(-k_1 \cdot [N_2\text{ph-F}] \cdot t) + A_2 \exp(-k_2 \cdot [N_2\text{ph-F}] \cdot t)$. The *inset* in **f** represents the rate constants from the second exponential obtained from **d** (k_2), plotted against the corresponding rate constants for dinitrophenylation of Lys *a*, as inferred from measuring H_2DIDS binding capacity of the modified cells

disulfonates (*Ramjeesingh* et al. 1981) and phenylsulfonates are known to bind (*Mawby* and *Findlay* 1982).

Although this residue is involved in the covalent reaction with stilbene isothiocyanates, it is not required for noncovalent H_2DIDS binding: only when several additional lysine residues are modified is the noncovalent

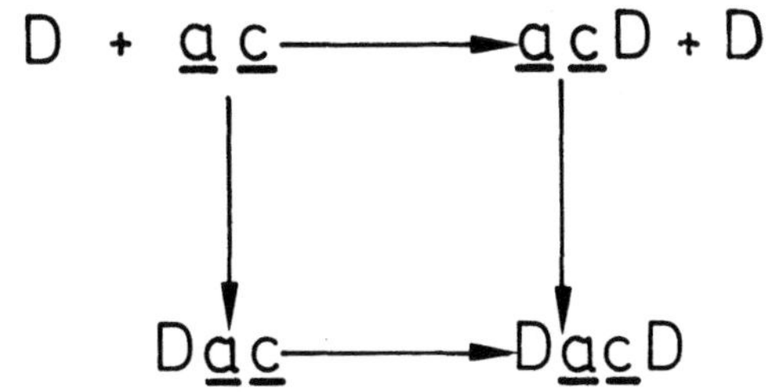

Fig. 23. Reaction diagram to describe the dinitrophenylation of Lys *a* and Lys *c* on the band 3 protein. After combination of N_2ph-F (*D*) with Lys *c*, the rate of transport is increased, and the rate of dinitrophenylation of Lys *a* is reduced. Inhibition of transport is also brought about when the dinitrophenylation of Lys *a* takes place in a band 3 molecule in which Lys *c* is already modified. (*Passow* et al. 1984b)

binding also abolished (*Passow* 1978; *Passow* et al. 1980a; *Schnell* et al. 1983). These observations are in accord with others that demonstrate the absence of effects on reversible stilbene disulfonate binding after changing the degree of protonation of Lys *a* by variation of pH (unpublished observations from our laboratory, and *Wieth* and *Bjerrum* 1983) or by reductive methylation (*Jennings* 1982a; *Jennings* and *Nicknish* 1984).

At N_2ph-F concentrations in the medium that cause up to 80% inhibition, about equimolar quantities of N_2ph residues become incorporated into the 17-kDa and the 35-kDa fragments of the band 3 protein. As in the 17-kDa segment, all N_2ph-F binding to the 35-kDa segment is to a lysine residue, which will be called Lys *c*. This second lysine residue is not identical to Lys *b*, the residue involved in intramolecular cross-linking by H_2DIDS. The rate of dinitrophenylation of Lys *a* but not of Lys *c* is reduced by the presence of stilbene disulfonates like DAS (*Passow* 1978) or DNDS (*Rudloff* et al. 1983).

Both Lys *a* and Lys *c* are related to the transport function of the band 3 protein and allosterically linked to each other. Dinitrophenylation of Lys *a* leads to an inhibition of transport, regardless of whether or not Lys *c* is also dinitrophenylated. Dinitrophenylation of Lys *c* reduces the rate of reaction of Lys *a* with N_2ph-F and slightly (1.4 times) enhances the transport of both monovalent and divalent anions (see Fig. 23).

The conclusions described above are based in part on experiments of the type presented in Fig. 22. The residual sulfate flux and the residual fraction of Lys *a* residues are plotted as measured after exposure of red cell ghosts to a range of N_2ph-F concentrations for a fixed length of time. For a first-order reaction, one would expect to find a simple exponential relationship between either one of these quantities and the N_2ph-F concentration to which the ghosts had been exposed during the dinitrophenylation period. A single exponential, however, does not provide an adequate fit. Instead, the sum of two exponentials is required, as expected for the modification of two interacting sites (see the drawn lines in Figs. 22, 24).

When the dinitrophenylation is carried out in a chloride medium (Fig. 22a,b) the influence of the second exponential becomes significant only at high degrees of dinitrophenylation. This is due to the fact that Lys *a* is more rapidly dinitrophenylated

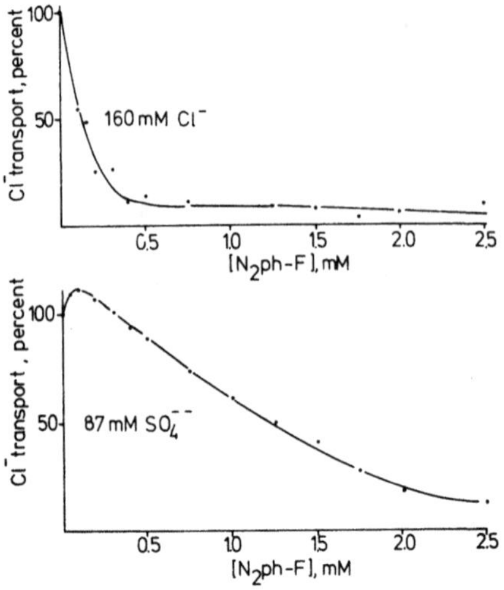

Fig. 24. Cl^- equilibrium exchange measured in red cell ghosts that had been dinitrophenylated at the N_2ph-F concentrations indicated on the ***abscissa*** in media containing 20 mmol liter^{-1} EDTA at pH 7.4 and either 160 mmol liter^{-1} NaCl plus 1 mmol liter^{-1} Na_2SO_4 or 87 mmol liter^{-1} Na_2SO_4. Flux measurements at 0°C after 30 min of dinitrophenylation at 37°C. (*Grygorczyk* and *Passow*, unpublished work)

than Lys *c*. Thus, there is only a small proportion of band 3 molecules in which dinitrophenylation of Lys *c* precedes that of Lys *a*. Consequently, there is little enhancement of transport and little retardation of the dinitrophenylation of Lys *a*.

If, however, the red cell ghosts are dinitrophenylated under otherwise identical conditions in sulfate media (Fig. 22d,e), the rate of dinitrophenylation of Lys *a* is reduced to about 1/20 the value in Cl^- media (compare Fig. 22b, and 22e) while the rate of dinitrophenylation of Lys *c* remains unaltered. As a consequence, the rate of dinitrophenylation of Lys ***c relative*** to Lys *a* is increased. Besides band 3 molecules in which Lys *a* residues are modified and hence transport is inhibited, there now occurs a considerable number of molecules in which Lys *c* is dinitrophenylated while Lys *a* is not yet affected. These latter molecules accomplish anion transport at a higher rate than the molecules in which neither Lys *a* nor Lys *b* had been altered. As a consequence, a situation arises where the decrease of the flux associated with the increase of band 3 molecules in which Lys *a* and Lys *a* plus Lys *c* are modified is more than compensated by the increase of the flux in those band 3 molecules in which only Lys *c* had been dinitrophenylated. The flux passes through a maximum (Fig. 22d; Fig. 24, lower panel). At higher degrees of dinitrophenylation even in those band 3 molecules in which Lys *c* was dinitrophenylated prior to Lys *a*, Lys *a* will also become dinitrophenylated, although at a reduced rate. This leads to inhibition of transport (Fig. 24, lower panel, see the reaction diagram in Fig. 23).

The different effects of dinitrophenylation of Lys *c* on anion transport and on the rate of dinitrophenylation of Lys *a* result in a nonlinear relationship between flux and number of unmodified Lys *a* residues (Fig. 22f). From the two exponentials that fit the data (see the drawn curves in Fig. 22d,e), it is possible to derive the rate constants for the modification of Lys *a* in molecules in which Lys *c* had already been modified. These constants can be obtained from the exponential that pertains to the time course of modification of Lys *a*. They can be plotted against the rate constants for modification of transport. This should yield a linear relationship, as was actually observed (inset Fig. 22f).

The effects of SO_4^{2-} and Cl^- show that the reactivity of Lys *a* depends on the nature of the substrate that is present during exposure to N_2ph-F. Less-detailed studies with other substrates yielded the sequence: $F^- \approx Cl^- > Br^- > I^- \approx SO_4^{2-}$ (130 mmol liter^{-1}, pH 7.4, 37°C). When no substrate is present, the rate of dinitrophenylation is somewhat higher than in SO_4^{2-} medium. Hence the addition of SO_4^{2-} causes a slight reduction of the rate of dinitrophenylation as compared with a substrate-free control, while the addition of Cl^- causes a considerable enhancement. This latter effect shows clearly that Lys *a* does not directly participate in complex formation with the substrate (*Passow* et al. 1980a).

In addition to substrates, inhibitors of anion transport also influence the reactivity of Lys *a*. Before discussing the effects observed, it should be noted that after completion of the reaction with Lys *a* the structure and function of a given band 3 molecule is irreversibly modified. However, the irreversible reaction of N_2ph-F with lysine residues is not preceded by reversible binding, which could induce a structural change of the band 3 protein molecule. *Hence, the rate at which the reaction takes place only depends on the structure of the native protein as it existed prior to the covalent reaction.* The rate of dinitrophenylation represents, therefore, an indicator of the conformational state of the transport protein and can be used to monitor the effect of substrates and inhibitors on the otherwise unperturbed transport protein.

When anion transport is blocked by inhibitors, the transport protein may become arrested in a specific conformeric state. Using a variety of different inhibitors it was explored in which configuration the arrest occurred: with Lys *a* exposed or buried (*Passow* et al. 1980a). The inhibitors could be classified into three different groups. One group of inhibitors prevented the access to Lys *a* presumably by sterical hindrance in the diffusive pathway and hence were not suited for conformation probing by measuring the rate of dinitrophenylation. These inhibitors included the stilbene disulfonates and *external* APMB, which are known to bind at or near Lys *a*. A second group of inhibitors, including phlorizin and positively charged furosemide derivatives, arrested the transport with Lys *a* exposed. The third group produced inhibition with Lys *a* buried. This group comprised *internal* APMB and Ca^{2+} which both exert the inhibitory influence by binding at the inner membrane surface.

In these studies (*Passow* et al. 1980a; *Passow* and *Fasold* 1980), the band 3 conformation with Lys *a* exposed was designated "cis," with Lys *a* buried "trans." These expressions alluded to the possibility that the two states of Lys *a* are representative of the outward- and inward-facing conformation of the band 3 protein as implied in the Ping-Pong model. As has already been mentioned on p. 112f., this could be expected if the affinities of chloride to the inward- and outward-facing substrate binding site were different (i.e. $K_{101} \neq K_{11}$). Different affinities were postulated because it seemed unlikely that in an asymmetrical structure like the band 3 protein, equal affinities could occur for the two orientations of the site. Allosteric effects of chloride binding, which could also affect the reactivity of Lys *a* were disregarded. It is clear, however, that this represents an oversimplification since allosteric effects could modulate the effects of recruitment associated with different affinities of the substrate to the two conformations of its binding site. Further work will be required to assess the role of allosteric effects.

In summary, the work on dinitrophenylation shows that Lys *a* at the stilbene disulfonate binding site may exist in two different states with greatly different reactivites. The equilibrium between the two states is influenced by substrates and inhibitors. The residue is allosterically linked to Lys *c*, which resides in the 35-kDa segment and whose accessibility to N_2ph-F is independent of the presence or absence of stilbene disulfonates and substrates.

4.3.1.3 Pyridoxal 5-Phosphate

Treatment with pyridoxal 5-phosphate (P5P) leads to an inhibition of anion transport, including the transport of the inhibitor itself, which is a substrate of the anion transport protein. Inhibition becomes irreversible after reduction with $NaBH_4$. Using tritiated $NaBH_4$, it is possible to label the site of P5P binding. Much of the binding to band 3 can be prevented by treatment of the cell membrane with H_2DIDS or DNDS. However, not all of the binding can be prevented, indicating that some of it takes place outside the H_2DIDS binding site (*Cabantchik* et al. 1975). A recent analysis (*Nanri* et al. 1983) revealed that DNDS only prevents binding to the 35-kDa chymotryptic fragment but not to the 60-kDa fragment. Moreover, it was possible to show that the binding to the former correlated with the inhibition of anion transport while the binding to the latter is ineffective. The site of action could be further narrowed down to the papain sensitive portion of the 35-kDa fragment (*Matsuyama* et al. 1983) and it is attractive to think that it is identical to Lys *b*.

4.3.1.4 Phenylisothiocyanate

Arylisothiocyanates have been widely used for the study of structure and function of the band 3 protein. As has already been mentioned, the hydrophilic sulfophenylisothiocyanate (IBS) reacts at the stilbene disulfonate binding site (*Ho* and *Guidotti* 1975; *Zaki* et al. 1975; *Reithmeier* and *Rao* 1979), presumably with Lys *a*. Arylisothiocyanates are able to react with nucleophilic groups of amino acids in the absence of water. Hence hydrophobic arylisothiocyanates can be used for labeling of intramembraneous portions of integral membrane proteins (for review, see *Sigrist* and *Zahler* 1982).

The hydrophobic phenylisothiocyanate inhibits anion transport (*Sigrist* et al. 1980). It combines with lysine residues in the hydrophobic core of the anion transport protein but not with Lys *a*. When three moles of phenylisothiocyanate are bound per mole of band 3, anion transport is completely inhibited. However, when Lys *a* is first blocked by sulfophenylisothiocyanate, then the subsequent binding of phenylisothiocyanate is reduced by one mole. The binding site that becomes inaccessible could be identified as Lys 42 on a 10-kDa piece of the 35-kDa segment (called Lys *d* in Table 4). This finding indicates an allosterical relationship between Lys *a* and Lys *d* (*Brock* et al. 1983).

Table 4. Allosterical relationships within the band 3 protein

Amino acid residue	Location Segment	Facing	Modification by	Allosterically linked to	Comments
1 Lys *a*	17-kDa	Out	Nph-F(1,2) Isothiocyanates (2,3) Reductive methylation (3a)	Transfer site internal Cl binding site (4) Lys *c*	Inhibition of transport upon dinitrophenylation (1,2) Modification by dinitrophenylation unaffected by external Cl^-, enhanced by internal Cl (4)
2 Lys *b*	35-kDa	Out	H_2DIDS (5)	Transfer site	Lys *a* and Lys *b* are cross-linked by H_2DIDS (5). Transport inhibited upon binding of P5P (6), methylation (3a)
3 Lys *c*	35-kDa	?	Nph-F	Transfer site Lys *a*	Accelerates transport of Cl and SO_4 upon dinitrophenylation (7). After dinitrophenylation of Lys *c* the rate of dinitrophenylation of Lys *a* is reduced (7)
4 Lys *d*	35-kDa	Inside hydrophobic region	PI	Transfer site Lys *a*	Transport inhibited, reactivity of Lys *a* to isothiocyanate reduced (3)
5 Lys *e*	35-kDa	Inside hydrophobic region	PI	–	Binding produces no effect on transport (3)
6 Lys *f*	35-kDa	Inside hydrophobic region	PI	–	Binding produces no effect on transport (3)

Table 4 (continued)

Amino acid residue	Location Segment	Facing	Modification by	Allosterically linked to	Comments
7 Carboxyl group	5–10-kDa papain fragment	Out	Carbodiimides	Transfer site	Inhibits transport, possibly without affecting anion binding to transfer site. Prevents H_2DIDS binding (8, 9, 10)
8 Dans *a*[a]	Location unknown		Dansylchloride	Transfer site	In absence of APMB accessible to dansylchloride (11) but not to PENS-Cl (12)
9 Dans *b*[a]	Location unknown		Dansylchloride	Transfer site	In presence of APMB accessible to dansylchloride (13) but not to PENS-Cl (12)
10 Dans *c*[a]		Out	Dansylchloride	Transfer site	In presence of APMB accessible to both dansylchloride and PENS-Cl (12)
11 Arg *a*	?	Out	Phenylglyoxal	Transfer site	Constituent of a substrate-binding modifier site (14), presumably distinct from Dalmark's (15) site. r_2?
12 Arg *b*	35-kDa	Out	Phenylglyoxal		Located near transfer site but cannot be recruited (14). r_1?
13 Arg *c*	?	In	Phenylglyoxal	Transfer site	Inhibits upon phenylglyoxalation (9)
14 Outward-facing substrate binding site[b]	?	Out	Phenylglyoxal NAP-taurine	Transfer site	Arg *a* is one of its constituents (14)
15 Inward-facing substrate binding site[b] (*Dalmark*, 15)	?	In		Transfer site	Inhibition of transport (16)

16 Ca binding site[b]	?	In	Ca	Transfer site, Lys *a*	Transport inhibited (17, 18). Lys *a* buried (19)
17 APMB binding site[b]	?	In	APMB	Transfer site Lys *a*	Tranport inhibited. Lys *a* buried (19)
18 Phlorizin binding site[b]	?	Out	Phlorizin	Transfer site Lys *a*	Transport inhibited. Lys *a* exposed (19)
19 Substrate binding site, different from site 15	?	In	NAP-taurine		H_2DIDS binding to outside prevents covalent NAP-taurine binding inside (20)
20 SH?	17-kDa	Inside the protein	pCMBS	Stilbene disulfonate binding site	pCMBS does not inhibit anion transport but binding of DBDS (21)

APMB, 2-(4′-aminophenyl-6-methylbenzenethiazol)-3′7-disulfonate; Dans *a*, dans *b*, dans *c*, dansylated, but as yet unidentified, amino acid residues; H_2DIDS, 4,4′-diisothiocyano dihydrostilbene-2,2′-disulfonate; N_2ph-F, 1-fluoro 2,4-dinitrophenol; P5P, pyridoxal phosphate; PI, phenylisothiocyanate; PENS-Cl, 2-(*N*-piperidine)-ethylamine-1-naphthyl-5-sulfonylchloride; pCMBS, *p*-chloromercurisulfonate; DBDS, see Fig. 14

[a] Band 3 not yet established as site of action
[b] The term "site" refers to a region of the transport protein comprising an unknown number of as yet unidentified amino acid residues that may be contributed by different adjacent segments of the peptide chain

References: (1) *Passow* et al. (1975); (2) *Zaki* et al. (1975); (3) *Kempf* et al. (1981); (3a) *Jennings* (1982a); (4) *Lepke* and *Passow* (1984), unpublished; (5) *Jennings* and *Passow* (1979); (6) *Nanri* et al. (1983); (7) *Rudloff* et al. (1983); (8) *Andersen* et al. (1983); (9) *Bjerrum* (1983); (10) *Wieth* et al. (1982); (11) *Legrum* et al. (1980); (12) *Raida* and *Passow* (1985); (13) *Lepke* and *Passow* (1982); (14) *Wieth* and *Bjerrum* (1982); (15) *Dalmark* (1976); (16) *Knauf* and *Mann* (1984a); (17) *Low* (1978); (18) *Gunn* et al. (1979); (19) *Passow* et al. 1980a); (20) *Grinstein* et al. (1979); (21) *Lukacovic* et al. (1984b)

4.3.2 Arginine Residues

Two reagents that inhibit anion equilibrium exchange are known to react selectively with arginyl residues in proteins: 1,2-cyclohexanedione (*Zaki* 1981) and phenylglyoxal (PG) (*Zaki* 1982, 1983, 1984; *Bjerrum* et al. 1983; *Wieth* et al. 1982a,c). At pH 10.0, 38°C, the time course of inhibition by PG is of the first order; the apparent rate constant of the reaction increases linearly with PG concentration and thus yields a concentration-independent second-order rate coefficient. Maximal inhibition is about 90% (*Wieth* et al. 1982a), but after repetition of the exposure to the reagent inhibition increases up to 98% (*Bjerrum* et al. 1983). The relationship between inhibition and binding is linear. At maximal inhibition nearly two PG molecules are bound per band 3 molecule corresponding to the modification of one arginyl residue per transport molecule. This residue resides on the 35-kDa fragment (*Bjerrum* et al. 1983). Inhibition by phenylglyoxalation is accompanied by the release of a chloride ion from the stilbene disulfonate binding site, in accordance with the view that this site encompasses the substrate binding site (*Falke* and *Chan* 1984).

The results described above were obtained in media of low substrate concentration. Increasing the Cl^- concentration or adding the stilbene disulfonate DNDS reduces the rate of reaction and alters the stoichiometry of the relationship between inhibition and binding. At 165 mmol $liter^{-1}$ KCl or at 2 mmol $liter^{-1}$ DNDS in the medium, the relationship is still linear, but inhibition becomes maximal only when four PG molecules are bound per band 3 molecule rather than the two observed in the absence of Cl^-. The stoichiometry of 4:1 is also seen when both 165 mmol $liter^{-1}$ KCl and 2.0 mmol $liter^{-1}$ DNDS are present, suggesting that the effects of the substrate and the DNDS take place at sites that overlap or are allosterically linked. An involvement of the stilbene disulfonate site can also be inferred from measurement of H_2DIDS binding to maximally phenylglyoxalated red cells: The covalent binding takes place more slowly, and much higher concentrations than in untreated red cells are required to achieve complete occupancy of the 10^6 H_2DIDS binding sites per cell. Such behavior would be expected if the phenylglyoxylation reduced the capacity for reversible H_2DIDS binding, which precedes the covalent bond formation. A reduction of reversible stilbene-disulfonate binding could actually be demonstrated, using the fluorescent DBDS as a probe: After treatment with PG, DBDS binding is reduced in proportion to the PG-induced inhibition of anion equilibrium exchange (Fig. 25). Thus, the modified arginine residue is indeed either directly or allosterically involved in the noncovalent binding of stilbene disulfonates.

Wieth et al. (1982a) have investigated the effects of a number of reversibly binding inhibitors of anion transport on the rate of inactivation by

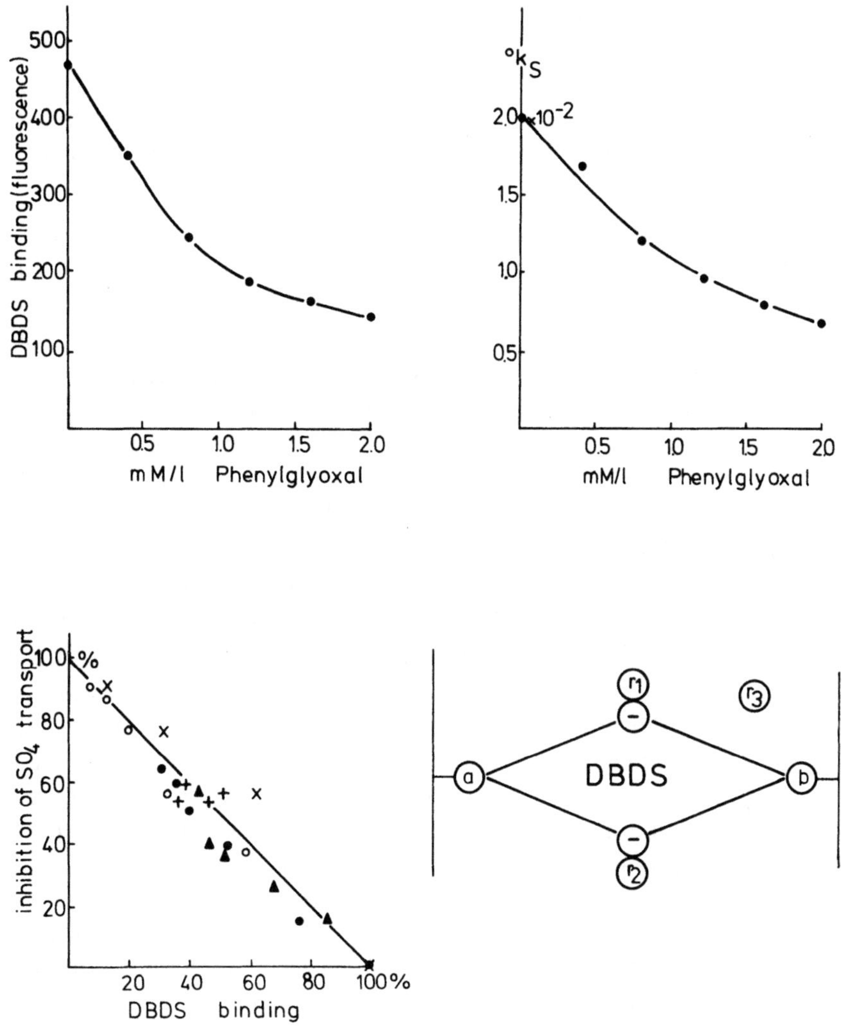

Fig. 25. Effect of phenylglyoxalation of the red cell membrane on DBDS binding to the band 3 protein and the rate of sulfate equilibrium exchange. Red cell ghosts had been phenylglyoxalated at a range of phenylglyoxal concentrations. Each batch of modified ghosts was subdivided into two, one for measuring DBDS binding (by fluorescence) the other for measuring transport. DBDS binding represents the difference between binding in the absence and presence of a large excess of H_2DIDS

phenylglyoxalation of the anion transport protein. They observed that among other inhibitors, and in contrast to DNDS, phloretin and niflumic acid do not affect the rate of inactivation. This finding could be accommodated in the hypothetical picture of the stilbene disulfonate binding site presented in Figs. 16 and 20. It would suggest that the positively charged site r_3, which supposedly combines with the negatively charged phloretin or niflumate, cannot be responsible for the inhibition of transport by phenylglyoxalation. Instead, inhibition is likely to be due to phenylglyoxalation of one of the other two positively charged groups (r_1 or r_2) that are allegedly involved in DNDS binding. These groups seem to remain susceptible to PG even after inhibitor binding to the positive charge r_3 near Lys *b* in the 35-kDa fragment of the band 3 protein.

The effects of Cl^- on the rate of inhibition by phenylglyoxalation could be easily explained if the modified arginine residue were part of the substrate binding site. If so, one could expect that the efficiency of Cl^- should be reduced by recruitment of the transport protein from outward-facing to inward-facing. In the low-Cl^- media, a large outward-directed Cl^- gradient exists, which recruits most of the transport protein molecules into the outward-facing configuration. Reducing the gradient should favor the inward-facing configuration. This can be monitored by measuring reversible stilbene-disulfonate binding and has actually been observed in untreated red blood cells. The rate and nature of the reaction of PG with the band 3 protein does not change, however, when the chloride concentrations inside are made equal to the low chloride concentration outside. The rate of reaction is independent of the Cl^- gradient across the membrane and entirely determined by the external Cl^- concentration. Thus, there is no evidence for a transport-related change of the accessibility of the PG-susceptible arginine residue.[10] This has led to an alternative explanation of the Cl^- effect. If one stipulates that the PG-sensitive arginine residue is located at the H_2DIDS binding site, then one can expect it to be one of the constituents of a cluster of positively charged groups (see Fig. 20). According to the fixed-charge hypothesis (see *Passow* 1969), such a cluster would give rise to a *local* positive surface potential with a consequent accumulation of diffusable, negatively charged counter ions. This would include not only Cl^- but also OH^- such that

$$Cl^-_{surface}/Cl^-_{bulk} = OH^-_{surface}/OH^-_{bulk} = \exp(-F\Delta\psi/RT)$$

(the symbols have their usual meaning). Thus, the pH in the surface would be different from that in the medium and would be determined by the chloride distribution ratio. Increasing the Cl^- concentration in the bulk phase at constant pH would decrease the pH in the vicinity of the PG-sen-

10 Even though the arginine residue titrated at about pH 12 (pp. 119, 123) cannot be recruited, *Wieth* and *Bjerrum* (1982), 1983) designate this residue "transfer site" (see above, p. 132, in connection with DNDS binding). Obviously, they employ this term in a broader sense than used here and in most of the literature, meaning that the site not only encompasses side chains involved in substrate binding but also groups that direct the substrate to the binding site without being directly involved in actual bond formation (approach sites)

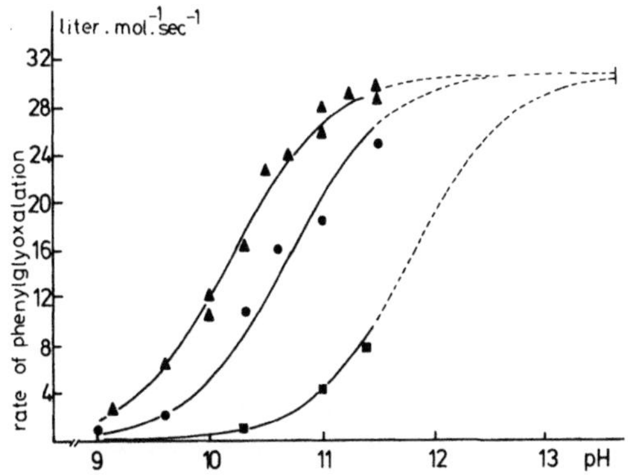

Fig. 26. pH dependence of the rate of irreversible inhibition of Cl^- equilibrium exchange by phenylglyoxalation. The apparent first-order rate coefficients for inactivation (*ordinate*) were measured as functions of extracellular pH (*abscissa*) at three different Cl^- concentrations in the medium: *Triangles, circles,* and *squares* refer, respectively, to 2, 8, and 165 mmol liter^{-1} Cl^- in the extracellular medium. 25°C. Titration curves were fitted to the experimental values, with single p*K* values of 10.2, 10.7, and 11.8 and a common maximal rate of reaction of 31 liters mol^{-1} s^{-1}. (*Wieth* et al. 1982c)

sitive arginine residue. Since the reaction with PG only takes place with the deprotonated form of arginine residues, this would reduce the rate of reaction, as was actually observed (*Wieth* et al. 1982a).

The pH dependence of the rate of modification of the anion transport system by phenylglyoxalation is compatible with a single apparent p*K* value of 10.6 at 5 mmol liter^{-1} Cl^- in the medium. This value gradually increases up to 11.7 when the Cl^- concentration is increased up to 160 mmol liter^{-1} (Figs. 26, 27a). Previous measurements of the pH depen-

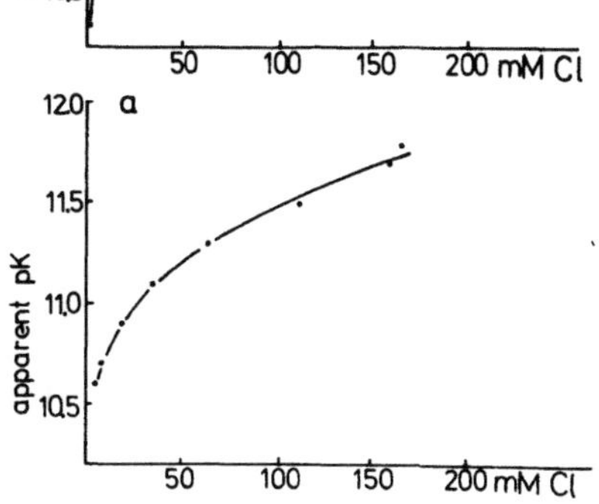

Fig. 27. a Apparent p*K* values derived from measurements of the pH dependence of the rate of phenylglyoxalation at varying Cl^- concentrations in the medium. 25°C. (*Wieth* et al. 1982c). **b** Apparent p*K* values of a transport-related, titratable group of the band 3 protein exposed to the outer cell surface (presumably the arginine residue r_1 in Figs. 16 and 20, or Arg *b* in Table 4) as a function of the Cl^- concentration in the medium; 0°C. This curve and the curve in **a** presumably refer to the same arginine residue. (*Wieth* and *Bjerrum* 1982)

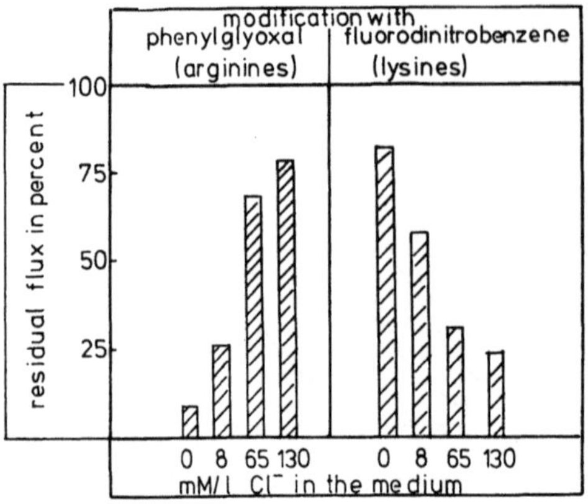

Fig. 28. Inverse effects of Cl^- on the rates of inhibition of anion transport by phenylglyoxalation and dinitrophenylation. Cl^- was varied at equal concentrations inside and outside. 37°C, human red cell ghosts. Data on phenylglyoxalation from *Zaki* (1982) and on dinitrophenylation from *Passow* et al. (1980a)

dence of Cl^- equilibrium exchange in untreated red cells can be described by dissociation curves with single apparent pK values that rise from 11 at 15 mmol liter^{-1} Cl^- in the medium to nearly 12 at 160 mmol liter^{-1} (Fig. 27b). The similarity of the effects of Cl^- on the pH dependence of the modification reaction and of the transport process leaves little doubt that the phenylglyoxalated group is identical with the group that controls the pH dependence of anion flux at high pH. Here again, the authors (*Wieth* and *Bjerrum* 1982) believe that the Cl^- effect does not repesent a change of the reactivity due to recruitment, but an independent effect by a change of surface potential and surface pH.

In this context it should be pointed out that, unlike the substrate Cl^-, the substrate SO_4^{2-} does not reduce the rate of phenylglyoxalation at high pH (*Zaki* 1982; *Wieth* et al. 1982a). Since the divalent SO_4^{2-} ion would be even more effectively accumulated near positively charged groups in the surface than the monovalent Cl^- ions, one would expect that they should exert a bigger effect on surface pH than Cl^-. Contrary to this expectation, in the pH range where the effect of Cl^- had been studied, the rate of phenylglyoxalation is independent of SO_4^{2-} concentration (*Zaki* 1982; *Wieth* et al. 1982a). Moreover, at low pH, where the probability of simultaneous binding of SO_4^{2-} and H^+ is increased, SO_4^{2-} conveys protection similar to Cl^- against phenylglyoxalate inactivation. Thus, protection against phenylglyoxalation seems to be afforded only when the complex of transport protein and divalent anion has been transformed by proton binding into a conformational state in which the transition from outward-facing to inward-facing becomes feasible (*Zaki* 1981).

Finally, it should be recalled that the lysine residue *a* which is involved in the covalent binding of N_2ph-F to the 17-kDa fragment of band 3 should be located within the same cluster of positive surface charges as the arginine residue discussed here. In terms of the surface potential hypothesis, increasing the Cl^- concentration should reduce the rate of dinitrophenylation of this residue similar to the reduction of the rate of phenylglyoxylation of the arginine residue. However, the reverse is observed (*Passow* et al. 1980b). It would seem attractive to believe, therefore, that Cl^- (or HSO_4^-) binding leads to reduction of the reactivity of the arginine residue and an enhancement of the reactivity of Lys *a*, while the binding of the substrate would reverse the respective reactivities (Fig. 28). Thus, the two residues could possibly be coupled with one another and their respective reactivities could be reflections of the occupancy of the transfer site. Both reactivities should be modulated by the surface potentials, as discussed by *Wieth* and his associates.

4.3.3 Carboxyl Groups

The titration data of *Wieth* and *Bjerrum* (1982) discussed in a preceding section showed that decreasing the extracellular pH at constant intracellular pH leads to an inactivation of Cl^- self-exchange with a pK of 5.2–5.4 (165 mmol $liter^{-1}$ extracellular Cl^-, 0°C). Between 0° and 10°C the ionization enthalpy is close to zero, suggesting that the dissociable groups represent carboxyl groups. This inference was supported by experiments with carbodiimides (*Craik* and *Reithmeier* 1984), including a water-soluble, nonpenetrating carbodiimide derivative (1-ethyl-3-(4-azonia-4.4-dimethylpentyl)-carbodiimide) that carries a quaternary ammonium group (*Wieth* et al. 1982b; *Andersen* et al. 1983; *Bjerrum* 1983). The latter reagent produces irreversible inhibition with a biphasic time course. The rates of inactivation during the initial rapid phase and the subsequent slow phase differ by a factor of about 100 and the transition from fast to slow phase takes place when the inhibition of Cl^- equilibrium exchange reaches 50%. The rapid phase is accelerated twofold, the slow phase more than 100-fold. Upon addition of tyrosin ethyl ester, about one to two molecules of the ester become incorporated per band 3 protein molecule (*Bjerrum* 1983). They are located in the papain-digestible 5- to 10-kDa segment of the 55-kDa peptide (see Figs. 1, 2) and are released into the supernatant when the digestion with papain is carried out. The residual flux after papain digestion of the carbodiimide/tyrosin ethyl ester-treated red cells is equal to the residual flux in papain-treated red cells that had not been pretreated with the chemical agents. Since papain is known to leave the substrate binding site unaltered (*Jennings* and *Adams* 1981, see p. 135) this would suggest that the modification of the carboxyl group also leaves the substrate binding site intact but changes the rate of translocation of the bound substrate. Thus the carboxyl groups are most likely modifier sites that are allosterically involved in the maintenance of the penetration pathway in a functional state.

The inhibition of anion transport by carbodiimide is associated with a loss of the capacity of the band 3 protein to combine with stilbene disulfonates reversibly. Since carboxyl groups are unlikely to be involved directly in the binding of the negatively charged stilbene disulfonates one may stipulate that they participate allosterically in the preservation of the normal configuration of the stilbene disulfonate binding site.

4.4 Chemical Modification of Unidentified Amino Acid Residues

4.4.1 NAP-Taurine

The sulfonate NAP-taurine (*N*-(4-azido-2-nitrophenyl)-2-aminoethylsulfonate) is a photoactive anion that, upon illumination, forms a nitrene which is capable of reacting covalently with aliphatic or aromatic amino acid residues. It is accepted by the band 3 protein as a substrate which is transported and competitively inhibits the transport of other anion species. The competition kinetics are different at the inner and outer membrane surface. At the inner surface, the apparent K_I for zero chloride concentration is about 730 μmol liter^{-1}, which is close to the half saturation concentration for NAP-taurine transport as measured under comparable conditions. The apparent dissociation constant for Cl^- as calculated from NAP-taurine competition is about 36 mmol liter^{-1} and hence close to the value for chloride binding to the inward-directed transfer site. Thus, the data suggest that the NAP-taurine binding site at the inner membrane surface is the transfer site. At the outer membrane surface, however, the apparent K_I value for zero Cl^- concentration (20 μmol liter^{-1}) is lower and the calculated half saturation concentration for the competing Cl^- much higher than at the inner membrane surface. The latter result is not compatible with competition for the outward-directed transfer site. However, the details of the inhibition kinetics indicate some NAP-taurine binding to an additional site, which could be identical to the outward-directed transfer site. Thus, the data suggest that NAP-taurine causes inhibition at the outer membrane surface predominantly by binding to a substrate-binding modifier site (*Knauf* et al. 1978a,b; *Knauf* 1979). This site is obviously distinct from Dalmark's modifier site, which is located at the inner membrane surface. It is possibly identical to the outward-facing modifier site described on p. 119, where deprotonation leads to an enhancement of anion equilibrium exchange.

NAP-taurine binding to the external modifier site is enhanced by a recruitment of the band 3 protein into the conformation with outward-oriented transfer site. Table 3 (p. 133) shows that the recruitment of the transport protein by raising the internal Cl^- concentration at constant external Cl^- concentration increases the apparent affinity for NAP-taurine binding. The percentage change is quite close to the change of apparent affinity for reversible binding of H_2DIDS, indicating that both sites are equally affected by the recruitment (*Knauf* et al. 1980, 1984). This agrees with the observation (p. 141) that the binding sites for NAP-taurine and stilbene disulfonates overlap (*Macara* and *Cantley* 1981b).

The effect of recruitment on NAP-taurine binding to the inner membrane surface has not yet been studied to the same detail. Nevertheless, a striking transmembrane effect has been observed. When H_2DIDS is first covalently attached to the stilbene disulfonate binding site in the outer cell surface, then NAP-taurine binding to the binding site at the inner membrane surface is largely suppressed (*Grinstein* et al. 1979). This observation does not necessarily indicate the sequestration at the outer cell surface of the same amino acid residue that is involved in NAP-taurine binding at the inner cell surface. It is quite possible that the effect is allosteric, similar to that of H_2DIDS binding to the outer cell surface on hemoglobin binding to the inward-facing 42-kDa domain of band 3 (*Salhany* et al. 1980).

4.4.2 Eosine 5-Maleimide

Using eosine maleimide, *Macara* et al. (1983) specifically labeled the stilbene disulfonate binding site on the 17-kDa segment. This leads to an irreversible inhibition of anion transport (*Nigg* and *Cherry* 1979). The fluorescence of the bound eosine is quenched by CsCl, but only when the nonpenetrating cesium is present at the cytoplasmic side of the membrane. According to the authors, this implies that after reaction of the inhibitor with band 3 at the outer membrane surface, the chromophor of the inhibitor becomes accessible to the aqueous environment on the cytoplasmic side of the membrane. Since the eosine maleimide is covalently bound, this would mean that the same amino acid residue that had been exposed to the outside during maleimide binding turns toward the inner surface after the reaction.[11]

If the interpretation of these results is correct, they would have farreaching implications. They would rule out the operation of transport mechanisms with more than one gate in series, such as the zipper mechanism proposed by *Wieth* et al. (1982b) and *Brock* et al. (1983). These mechanisms would be unlikely to permit the same amino acid residue to become

11 Attempts to demonstrate the alternation of an amino acid residue across the rate-limiting barrier in the band 3 protein had first been made by *Grinstein* et al. (1979) and *Passow* and *Zaki* (1978). Although transmembrane effects could be demonstrated, it was not possible to rule out that they were due to allosterical interactions rather than to movements of one and the same amino acid residue

alternately exposed to both the outer and inner surface of the rate-limiting structure in the transmembrane channel. "Lock in" mechanisms of the type described by *Passow* et al. (1980b) and *Knauf* et al. (1980) would, however, be compatible with the findings described (see p. 180, Fig. 29).

4.4.3 Dansylation

As has already been stated in a preceding section, the band 3 protein-mediated exchange of monovalent and divalent anion species is inversely affected by the dansylation of the red cell membrane. The former is slightly inhibited, the latter enhanced. The enhancement is confined to the pH range above the pH maximum seen in untreated red cells, i.e., above about pH 6.3. The enhanced transport of the divalent anions can be inhibited by stilbene disulfonates.

The enhancement actually observed depends on both Cl^- concentration and pH that existed during exposure to dansyl chloride. It can only be brought about when Cl^- ions are present during dansylation. The chloride concentrations required to achieve maximal enhancement as measured under standard conditions are larger than necessary for substrate saturation of the transfer site. At physiological Cl^- concentrations, there exists a pH optimum around pH 6.6 (*Legrum* et al. 1980; *Lepke* and *Passow* 1982).

Dansylation at the optimum leaves the lysine residues at the stilbene disulfonate binding site untouched. Nevertheless, both the 17-kDa and 35-kDa segments of the band 3 protein become heavily labeled. For this reason, it has not been possible to localize the site of action yet. Nevertheless, we are entitled to believe that the enhancement of transport is related to the modification of at least three different amino acid residues. One of them is accessible to dansyl chloride in the presence of Cl^- alone. The two others become available when, in addition to the chloride, the nonpenetrating disulfonate APMB is present in the medium. Of the latter two residues, only one can be modified when the penetrating dansyl chloride is replaced by a nonpenetrating derivative, 2-(*N*-piperidine)-ethylamine-1-napthyl-5-sulfonyl-chloride, called PENS-Cl (*Raida* and *Passow* 1985).

It should be recalled that the potentiating agent APMB is an inhibitor of anion transport (*Zaki* et al. 1975). For this reason the flux measurements that were needed to determine its effect on the modification by dansylation of the anion transport system had to be performed after removal of the agent from the medium at the end of the dansylation period. In the same manner, the effects on dansylation of two other agents were studied which, like APMB, produce inhibition of anion transport by combination with the stilbene disulfonate binding site: DNDS and DAS. DNDS was found to prevent the effect of dansylation virtually completely. DAS exerted no visible effect as compared with a control that had been dansylated in the absence of additional agents. Thus, three closely related agents that combine with the same site on the band 3 protein caused three entirely different allosteric effects on the dansyl chloride binding sites (*Lepke* and *Passow* 1982).

Allosterical effects are usually reciprocal. To see whether this also applies to the interactions between the binding sites for dansyl chloride and stilbene disulfonates, red cells were first dansylated in the presence of a large excess of APMB. After removal of the APMB, determinations were made of the apparent K_I values for the inhibition by APMB, DAS, and DNDS of the enhanced anion equilibrium exchange in the modified cells. K_I of DNDS was unaltered, while the K_I's of DAS and APMB were reduced to 1/3 and 1/40, respectively, of the value in the unmodified control. This result shows that dansylation of the three sets of susceptible sites leads to an allosterical modification of the stilbene disulfonate binding site and an ensuing change of the specificity of the inhibition by three different inhibitors. Thus, the interactions between the binding sites for stilbene disulfonates and dansyl chloride are mutual indeed (*Legrum* et al. 1980; *Lepke* and *Passow* 1982).

The qualitatively different effects of three structurally similar inhibitors that act by combination with the same site are difficult to reconcile solely with differences of induced fits. It seems instructive, therefore, to consider two alternative hypotheses.

The first hypothesis follows up the consequence of the previous suggestion that the H_2DIDS binding site may exist in two different conformeric states, B and B^*. If the affinities to a given inhibitor S differed, then B and B^* would compete for S. The following reactions could occur: (1) $S + B \rightleftharpoons SB$, (2) $S + B^* \rightleftharpoons SB^*$, (3) $SB \rightleftharpoons SB^*$, and (4) $B \rightleftharpoons B^*$. If the forms B^* and SB^* are susceptible to dansylation (possibly with different rate constants), while the forms B and SB are not, then an inhibitor with a higher affinity to B than to B^* should reduce the rate of dansylation; it would act like DNDS. If the inhibitor has a higher affinity to B^* than to B, then it should expose additional dansyl chloride binding sites and thus behave like APMB. If the affinities are about equal, then there should be little if any effect, as observed with DAS. Conversely, dansylation could shift the equilibrium between B and B^* and thus account for changes of the apparent K_I values for the reversibly binding inhibitors.

The second hypothesis would stipulate that the stilbene disulfonate binding sites exist in a single conformational state which offers two different overlapping areas for binding of the same inhibitor molecule. Thus the ratio of the occupancies of these areas could be different for different inhibitors. This would lead to different effects on dansylation similar to the inhibitor binding to two different conformers discussed above. On the basis of the currently available information, the first hypothesis seems preferable, but the second cannot be ruled out. For this reason, at the present time, the two hypotheses may serve to illustrate a problem that is typical for inhibition at a complex site in a complex transport molecule in which complex allosteric relationships exist.

4.4.4 *Mercurials*

Like other SH reagents, the mercurial pCMBS (*p*-chloro mercuribenzene sulfonate) has no directly assessable effect on anion transport in human red cells. Nevertheless, *Solomon* and his associates have demonstrated in a series of papers (*Lukacovic* et al. 1984b; *Yoon* et al. 1984) that there exist at least allosterical relationships between a specific pCMBS binding site, possibly the SH group on the 17-kDa tryptic-chymotryptic transmembrane fragment of band 3, and the stilbene disulfonate binding site. With increasing pCMBS concentration in the medium, the apparent dissociation constant for DBDS binding is reduced while the number of available DBDS binding sites does not seem to be affected. This suggests strongly that there is no competition for a common binding site between the two agents, but that the pCMBS binding exerts an allosteric effect on the affinity of the stilbene disulfonate binding site. Moreover, the rate of the slow conformational change of the band 3 protein that follows upon initial complex formation between DBDS and the stilbene disulfonate binding site is reduced (*Lukacovic* et al. 1984b).

pCMBS and related compounds, like pCMB or $HgCl_2$, inhibit water transport. Other SH reagents like NEM, iodoacetate, or iodoacetamide produce no inhibition of water transport, but also do not interfere with DBDS binding. This and other observations prompted *Solomon* and his co-workers to propose that, perhaps, both water and anion transport are mediated by the band 3 protein. They derived further support for this view from the use of stilbene derivatives which, in place of the isothiocyanate groups of DIDS, contain -HgCl residues and thus are capable of reaction with SH groups. These compounds inhibit anion transport instantaneously and completely with low apparent K_I values and water transport slowly and only partially (*Yoon* et al. 1984).[12] Stilbene disulfonates do not affect water transport; thus *Yoon* et al. stipulate that the site of inhibition of water transport is distinct from the stilbene disulfonate binding site but nevertheless located in the same channel that serves anion transport. In this context, it may be recalled (see p. 94) that, on the basis of entirely different evidence, *Benz* et al. (1984) suggested that the band 3 tetramers may form the water channels.

The mercurials mentioned above also enhance passive cation efflux. In studies with liposomes containing purified band 3 protein, *Lukacovic* et al. (1984a) also found an enhancement of K^+ efflux. They suggest, therefore, that under suitable conditions band 3 may serve as a channel for anions, cations, and water.

12 Another SH reagent that has been reported to inhibit water transport in red cells (*Brown* et al. 1975) and found to block anion exchange (*Reithmeier* 1983) is 5,5′-dithio bis(2-nitrobenzoic acid)

4.5 Interactions of Protomers of Band 3 Oligomers. Inferences from Studies of Chemical Modification

The work with chemical modifiers shows that the distances between the H_2DIDS binding sites in adjacent protomers of the band 3 oligomers in the red cell membrane are small, that interactions between them may take place but that, nevertheless, each protomer seems to accomplish transport essentially as an independently operating unit.

The evidence for the close proximity of adjacent stilbene disulfonate binding sites comes from resonance energy transfer measurements and suggests a site-to-site distance of about 28–52 Å (*Rao* et al. 1979). Some of the bulkier stilbene disulfonate derivatives used for the energy transfer measurements occupy not only their own site but also part of the stilbene disulfonate binding site in the nearest neighbor, thus preventing the binding of another molecule to that site (*Macara* and *Cantley* 1981a). For the less bulky stilbene disulfonate H_2DIDS, neither overlap nor intersubunit interactions have been observed so far. The rate of intramolecular cross-linking by H_2DIDS is independent of the occupancy of neighboring H_2DIDS binding sites, suggesting that the rather spectacular conformational changes associated with H_2DIDS binding within each protomer (see p. 173ff.) do not affect the H_2DIDS binding site of the neighbor (*Kampmann* et al. 1982).

Nevertheless, in other instances, interactions have been observed. After phenylglyoxalation of a specific arginine residue in more than 90% of the band 3 protomers, the affinity and rate of H_2DIDS binding to about 50% of the protomers is greatly reduced. Presumably the chemical modification has changed the relationships between the H_2DIDS binding sites in adjacent protomers such that now the binding of H_2DIDS to one of these affects the binding to the other (*Zaki* 1981; *Bjerrum* et al. 1983). Carbodiimide rapidly inactivates 50% of the band 3 molecules. Further inactivation proceeds much more slowly, suggesting that the modifications of one protomer reduces the rate of modification of any other. In spite of these interactions, the capacity for H_2DIDS binding to the unmodified protomers is preserved while it is lost in the modified protomers. Thus the interactions seem to be confined to carbodiimide binding sites in different protomers, and to the binding sites of carbodiimide and H_2DIDS in the same protomer (*Andersen* et al. 1983; *Bjerrum* 1983).

Regardless of the intersubunit interactions described, no evidence has been obtained for cooperation of two or more protomers in anion transport. The relationship between the binding of H_2DIDS or its derivatives and the inhibition of transport is always linear (*Lepke* et al. 1976 and many others), even when the bulky derivatives are used whose binding shows the negative cooperativity mentioned above (*Rao* et al. 1979). Moreover, regardless of the effect of phenylglyoxalation on reversible H_2 DIDS binding, the relationship between inhibition of transport and the binding of phenylglyoxal to its specific binding site is always linear over the whole range between 0% and 100% (*Bjerrum* et al. 1983). In this context it is pertinent to recall that the transport protein moves 10^6 anions per half cycle across the red cell membrane, corresponding to one anion per band 3 protomer (*Jennings* 1982b). Thus the currently available evidence is entirely compatible with the view that each protomer acts as an independently operating unit.

Even if each protomer acts independently, the possibility exists that the oligomeric state is nevertheless necessary to maintain each protomeric subunit in a functional state. This point has been stressed by *Jennings* (1984) in a topical review on the oligomeric state of the band 3 protein and its transport function. *Jennings* places particular emphasis on the work of *Boodhoo* and *Reithmeier* (1983), who attached band 3 dimers to a derivatized sepharose under conditions where only one of the protomers in each dimer became covalently bound to the sepharose. Each of these dimers was found to bind two stilbene disulfonate molecules. When, however, the noncovalently attached protomer of each dimer was removed by washing the column with a dissociating agent, the remaining monomer was unable to bind a stilbene disulfonate. Assuming that the method employed for the removal of the noncovalently attached protomer did not denature the covalently bound one, it was concluded that the interactions between the subunits were necessary to maintain the transport protein in a functional state.

4.6 Influence of the Composition of the Lipid Bilayer on the Band 3 Protein-Mediated Anion Transport

The activities of the transport systems for hexoses, monocarboxylates, and inorganic anions in the red blood cells of different mammalian species show marked differences. Attempts have been made to correlate them with differences of phospholipid composition and cholesterol content of the cell membranes (for a brief review, see *Deuticke* et al. 1980). Correlations have in fact been observed. They show some scatter, presumably because the transport rates are expressed per square centimeter of membrane surface rather than per transport molecule involved. Nevertheless, it is believed that the picture is dominated by lipid-related effects.

The rate of inorganic anion exchange is linearly correlated with the molar fractions of phosphatidyl choline (PC) and sphingomyeline (SM). The former correlation is positive, the latter negative (*Gruber* and *Deuticke* 1973; *Lu* and *Chow* 1982). Since these two choline-containing phospholipids replace each other in the red cell membranes of the various species, it remains open whether the increase of PC, the decrease of SM, or the ratio of the two is the essential variable involved.

Differences of the degree of unsaturation of the fatty acids of the phospholipids are also correlated with the rate of anion transport in the red cells of different species (*Deuticke* 1977; *Lu* and *Chow* 1982). Thus, a plot of transport rate against the 20:4 content is similar to that against PC, and a plot against the sum 18:1 plus 18:2 similar to that against SM. Consequently, the dependence of anion transport on the SM/PC ratio could reflect a dependence on fatty acid composition. However, this interpretation of the findings is not supported by experiments in which the fatty acid composition of rat erythrocyte membranes was varied by suitable diets (*Deuticke* et al. 1980). In these experiments, the ratio SM/PC as well as the rates of SO_4^{2-} equilibrium exchange remained unaltered. This lends support to the idea that the species differences depend on this ratio rather than on the degree of unsaturation of the fatty acids (*Deuticke* et al. 1980).

In addition to the phospholipids, the cholesterol content also affects the anion exchange. In human red cells, experimental incorporation or depletion of cholesterol is associated with decrease or increase, respectively, of anion exchange (*Grunze* et al. 1980). Similar changes were observed in red cells of patients who suffered from chronic cholestasis, which leads to a pathological increase of the red cell's cholesterol content (*Jackson* and *Morgan* 1982). The changes of anion transport are accompanied by changes of the arrangement of the band 3 protein in the lipid bilayer (*Borochov* et al. 1976) and its internal organization as inferred from studies of fluorescence quenching and polarization (*Borochov* and *Shinitzky* 1976; *Borochov* et al. 1976; *Klugerman* et al. 1984; *Shinitzky* and *Rivnay* 1977). In addition, it has been suggested that the aggregation of the band 3 protein may be increased (*Mühlebach* and *Cherry* 1982).

In order to furnish some information for a causal analysis of the observed correlations, *Deuticke* and his associates studied the effect of systematic variations of the major classes of phospholipids and cholesterol in reconstituted vesicles containing Triton X-100-extracted band 3 from human red cell membranes. They found that (1) in mixtures of PC and cholesterol, anion transport is barely affected by the cholesterol content of the membranes up to the highest cholesterol content used, viz., 27 mol% (*Köhne* et al. 1981); (2) in mixtures of SM and PC, containing approximately 25 mol% cholesterol, anion transport decreases considerably at high SM content; (3) the transport remains unaltered when PC is replaced by PE; and (4) in mixtures of PS and PE, anion transport decreases when the PS content exceeds about 30 mol% (*Köhne* et al. 1983).

The effects described are not directly comparable to those observed in intact red cells. In the vesicles, the phospholipid composition of outer and inner leaflet of the lipid bilayer should be rather similar. This is not the case in the intact red cells where the outward-facing leaflet contains predominantly PC and SM while the inward-facing leaflet contains most of the PE and all of the PS (*Zwaal* et al. 1973). The total PS content of the membranes of the red cells of all mammalian species studied so far is about 15%; hence, the molar fraction in the inner leaflet should be about 30%, i.e., close to the fraction that in the vesicles produces inhibition. The outer leaflet contains SM at molar fractions, varying between 20 and 100 mol% in the red cells of the various species studied, i.e., it covers a range where anion transport in the vesicles decreases strongly. Since variations of PC have little effect on transport in the vesicles, it seems that the rate of anion exchange in the cells is also essentially dependent on the molar fraction of SM rather than on the inverse changes of the molar fraction of PC. However, the band 3 protein has an asymmetrical structure; thus, the effects of the lipids on the protein's inward- and outward-facing portions need to be known separately before the influence of the lipids can be assessed more definitively.

In spite of these limitations, two comments on the possible cause of the relation between lipid composition and anion transport may be permissible.

1. Variations of lipid composition are likely to introduce changes of general physical properties of the membrane such as surface charge and fluidity. The PS should convey a negative charge to the inner membrane surface and the specific unsaturated and long chain fatty acids of the SM may affect the fluidity. An experimental study of the effects of surface charge has not yet been performed, and the dietary variations of the fatty acid composition of the phospholipids mentioned above do not seem to support the fluidity concept (*Deuticke* et al. 1980). However, experimental incorporation of cholesterol into membranes of red cells leads to both a decrease of fluidity and of anion transport (see discussion in: *Deuticke* et al., cited above).

2. It has been suggested that a specific interaction between cholesterol and the band 3 protein plays a decisive role in the control of anion transport (*Schubert* and *Boss* 1982). It has been hypothesized that there exists one inhibitory cholesterol binding site per band 3 protein molecule whose occupancy is dependent not only on the cholesterol content of the membrane but also on the SM/PC ratio in the bilayer. This hypothesis is based on work with lipid monolayers from which the occupancy of a high-affinity cholesterol binding site in the band 3 protein was inferred to be positively correlated with the cholesterol and SM content, and negatively correlated with the PC content of the monolayer. Thus the conditions that

favored an occupancy of the binding site corresponded to those associated with a decrease of anion transport in the red cell (*Schubert* and *Boss* 1982).

4.7 Stilbene Disulfonate Binding Site Revisited

The easy accessibility of the stilbene disulfonate binding site and the high susceptibility of at least several of its constituents to chemical modification have made this site a preferred object for biochemical and biophysical studies of the molecular mechanism of anion transport. Nevertheless, the considerable space allotted above to the discussion of this site does not only reflect the availability of many data. It is also a response to the putative significance of this site for anion transport. Several of its constituents play a role in substrate binding either directly or indirectly, perhaps by directing the anion toward a gate at the mouth of an anion channel. Other constituents of the site seem to be involved in maintaining the anion binding and transporting region in a functional state. It seems appropriate, therefore, to conclude the discussion of the transport-related molecular properties of the band 3 protein by a synopsis of the observations on the stilbene disulfonate binding site and of the inferences drawn thereof in the preceding sections. First, the properties of the site itself will be considered. Thereafter follows a summary of the allosterical relationships between some of its constituents and other regions of the transport molecule that seem to participate in transport-related conformational changes of the band 3 protein.

4.7.1 Structure and Function of the Site

In summing up, the stilbene disulfonate binding "site" is a region of the transport protein with a diameter of about 20 Å. It is easily accessible for stilbene disulfonates from the external medium, provided the transport protein is in its outward-facing conformation. It overlaps with the binding site for substrate anions and, in part, with the binding sites for other inhibitors. After treatment with external papain, a 5-kDa piece of the peptide chain is excised but not released. The capacity to bind stilbene disulfonates is retained. The transport-related conformational changes of the substrate binding site are altered. The site forms a hydrophobic niche with one or several tryptophane residues nearby, and with at least four, probably more, positively charged groups that are easily accessible from the outer medium and separated from the inner membrane surface by a distance of about 42–58 Å. All groups combined contribute to create a local electrostatic field that attracts anions, repels cations, and leads to the establishment of a surface pH that differs from that in the bulk phase. In addition, each group exhibits specific properties that may serve specific purposes for the transport process.

The individual properties of the various groups have been explored by acid-base titration and chemical modification, and attempts to assess their functional significance have been made by studying the influence of variations of substrate concentration on the binding of protons and the other modifiers.

Two of the positively charged groups could be identified as lysine residues. One of them is contributed by the 17-kDa segment of the peptide chain and resides in the first (*Ramjeesingh* et al. 1983) or second (*Tanner* et al. 1980) outward-facing loop, counted from the N-terminal end. Their p*K* values are abnormally low, which is presumably due to the positive charge of two to three neighboring arginine residues, one of which is located in the 35-kDa fragment.

Lys *a* and Lys *b* exhibit surprising specificities of their susceptibilities to covalent chemical modification by dansylation, dinitrophenylation, and reductive methylation. Neither one of the two residues can be dansylated under conditions where other regions of the band 3 protein become heavily labeled with dansyl chloride and divalent anion transport becomes accelerated. Lys *a* is more easily dinitrophenylated than Lys *b*, while the susceptibility of the two residues to reductive methylation is the inverse.

Increasing the chloride concentration enhances the dinitrophenylation of Lys *a*, indicating that it is not directly involved in substrate binding.

Both sulfophenylisothiocyanate and pyridoxal phosphate are substrates of the anion transport system. It is likely, therefore, that the noncovalent binding that precedes their covalent attachment requires the positive charges that are involved in reversible stilbene disulfonate binding. The covalent reaction of the two agents takes place at two different, as yet unidentified, amino acid residues. Sulfophenylisothiocyanate becomes attached to the 17-kDa segment (presumably to Lys *a*), pyridoxal phosphate to the 35-kDa segment (presumably to Lys *b*).

The substrate NAP-taurine binds upon illumination covalently to an unidentified amino acid residue on the 17-kDa segment. The binding site is also close to a chloride-binding site, or allosterically linked to it. This site is identical neither to the transfer site nor to another substrate binding modifier site, which is located at the inner membrane surface (Dalmark's modifier site). Upon deprotonation of this binding site, anion equilibrium exchange is enhanced. The deprotonation occurs with a p*K* of about 10.7, suggesting the involvement of an arginine residue in a positively charged environment. Assigning a location of this residue within the stilbene disulfonate binding site is further supported by this residue's capacity to become recruited by substrate gradients, similar to the recruitment of the transfer site.

The arginine residue mentioned in the preceding paragraph is distinct from another one that titrates around pH 12 and whose deprotonation leads to inhibition of anion transport. It has been proposed that the former constitutes part of the transfer site. However, since it cannot be recruited by substrate gradients, it seems that its role is confined to directing the substrate to the transfer site. It would thus exert a function similar to that of arginine residues of many of the enzymes with anionic substrates.

The reversibly binding DNDS no longer combines with the stilbene disulfonate binding site when this arginine residue is deprotonated. It has been proposed, therefore, that this residue, together with the other arginine residue mentioned above, is involved in the electrical neutralization of the negative charges of DNDS. This interpretation is, however, only tentative since the DNDS binding site, like the substrate binding site, can be recruited. This discrepancy casts a shade of doubt on the hypothesis that the arginine that titrates at pH 12 is indeed a constituent of the stilbene disulfonate binding site and not an allosterically linked site.

The functionally significant amino groups that are modified by phenylisothiocyanate, the carboxyl group that is modified by carbodiimides and the unidentified groups that react with dansylchloride are localized outside the positively charged area of the stilbene disulfonate binding site and are only allosterically linked to the maintenance of its function. They will be discussed in the context of the allosteric relationships between stilbene disulfonate binding site and other locations within the band 3 protein.

4.7.2 Allosteric Interactions with Functionally Important Sites of the Band 3 Protein

The binding and translocation of substrate anions by the band 3 protein is associated with changes of the arrangement of the peptide chain, which may result in alterations of the sensitivity to chemical modification of one or several amino acid residues that are involved in neither anion binding nor translocation and may be located far away from the transfer site. Conversely, chemical modification of these amino acid residues can be expected to give rise to changes in the size and rate of fluctuations of the energy barriers that determine the rate of anion transfer across the membrane. These allosteric interactions are an expression of the functional interrelationships between different portions of the large transport molecule. In the preceding sections, many examples have been described of such allosteric interactions between the transfer site and so-called modifier sites and among these modifier sites.

The study of the interactions between the various sites should enable one to establish a map of functional interrelationships between different portions of the transport protein. The information available is still inadequate for this ultimate goal but interesting observations have been made. Although overlap with the material presented in the preceding sections is inevitable, a brief summary of the hitherto-described interactions may be useful (Table 4, p. 155ff.).

1. Four different lysine residues have been identified which, when modified, produce a change of anion transport. Although two of them (Lys *a* and Lys *b*) are constituents of the stilbene disulfonate binding site, none of them seems to be directly involved in substrate binding to the transfer site. The nature and extent of the effect produced depends on the locations of the sites and the chemical nature of the agent used for modification. Dinitrophenylation of Lys *a* leads to inhibition, of Lys *c* to an enhancement of anion exchange. The inhibition that ensues as a consequence of the modification of Lys *a* is only complete when the modification is performed with N_2ph-F or certain isothiocyanates like isothiocyano phenylsulfonate, but remains incomplete after exhaustive reductive methylation. Allosterical interactions between Lys *a* (in the 17-kDa segment) on the one hand and either Lys *c* or Lys *d* (both in the 35-kDa segment) on the other hand have been demonstrated. Thus, after dinitrophenylation of Lys *c*, the rate of dinitrophenylation of Lys *a* is reduced; after thiocyanylation of Lys *a*, Lys *d* can no longer react with phenylisothiocyanate.

2. One or two carboxyl groups are allosterically involved in the maintenance of the functional state of the stilbene disulfonate binding site and the anion penetration pathway, although possibly without influencing the anion binding to the transfer site.

3. The dansylation of each of the three sites listed in Table 4 leads to a considerable enhancement of divalent anion transport with a concomitant slight inhibition of monovalent anion transport. Thus the effects are different from those of a modification of Lys *c* by dinitrophenylation, which slightly enhances both monovalent and divalent anion movements.

4. The self-inhibition of anion transport at high substrate concentrations indicates the existence of a modifier site that combines with the substrate and is allosterically linked to the transfer site. According to *Knauf* and *Mann* (1984a), the site (No. 15 in Table 4) is located at the inner membrane surface and is not identical with the NAP-taurine binding site at the outer membrane surface. However, there exists a second substrate binding site at the outer membrane surface. According to the Danish workers, the site (No. 14 in Table 4) is accessible to *external* H^+, phenylglyoxal, and overlaps with the external NAP-taurine binding site.

5. Several instances are known where sites that are allosterically linked to the stilbene disulfonate binding site reside on the opposite surface of the membrane: (1) intracellular Ca^{2+} and (2) intracellular APMB reduce the reactivity of Lys *a* at the extracellular surface (*Passow* et al. 1980a,b).

Binding of external H_2DIDS (3) renders an inward-facing chloride-binding site (presumably a constituent of the anion channel) inaccessible to covalent reaction with NAP-taurine and (4) increases the capacity of the 42-kDa domain of the band 3 protein to bind hemoglobin and glyceraldehyde phosphate dehydrogenase (*Salhany* et al. 1980).

In other instances allosterical linkages between the stilbene disulfonate binding site and modifier sites in the interior of the band 3 molecule have been found. Phenylisothiocyanate reacts with a lysine residue inside the hydrophobic portion of the 35-kDa segment, thereby blocking anion transport and reducing the reactivity of Lys *a*. pCMBS presumably reacts with an SH group on the 17-kDa transmembrane segment of band 3 without inhibiting anion transport, but nevertheless modifying the kinetics of stilbene disulfonate binding.

6. Finally, mention should be made that the arrest of anion transport by noncompetitively acting inhibitors may stabilize different conformeric states of the stilbene disulfonate binding site. For example, inhibition by external phlorizin and certain derivatives of furosemide (but not furosemide itself) is accompanied by an increased accessibility of Lys *a* for dinitrophenylation while inhibition by many other agents, including internal APMB and Ca^{2+}, is associated with a decreased accessibility of that residue (*Passow* et al. 1980a,b).

7. In addition to the intramolecular allosterical relationships, the lipid composition of the bilayer influences the conformation of the transport protein and its capacity to transport anions. Possibly this influence is exerted via a cholesterol-binding modifier site whose occupancy is not only a function of the molar fraction of cholesterol in the bilayer but also of the molar fractions of sphingomyelin and phosphatidylcholine.

The results described show that the combination of substrates or stilbene disulfonates with their respective binding sites causes allosteric effects that can be monitored at least at 14 distinct locations of the transport protein. This demonstrates most impressively that the transport protein is quite flexible, and that allosteric effects are transmitted over considerable distances between inner and outer membrane surface, and between predominantly hydrophilic and hydrophobic portions of the peptide chain. In some instances it was possible to establish correlations with substrate binding, but an interpretation in terms of the molecular mechanism of the transport process is still out of reach. Nevertheless it is clear that the enormous flexibility of the molecule would make it quite suitable to act as a channel with variable barrier heights.

5 Models of Anion Transport

The guiding lines for the experimental exploration of unknown mechanisms are provided by models. The kinetics of anion transport seem to be reconcilable with *Läuger* et al.'s (1980) model of a channel with variable-energy barriers. The model accounts for the occurrence of the large electrically "silent" anion exchange and the small diffusive anion flow. It provides, however, only a rather formalistic description of allosterical interactions and conformational changes within the large transport molecule. There remains, therefore, the task of identifying the structures and their changes that are responsible for the formation of the energy barriers and their fluctuations. The models of such structures may be subdivided into archaic and recent ones (Fig. 29).

The oldest model was that of *Mond* in 1927, who believed in the existence of water-filled pores, lined with positive fixed charges which permitted access to anions and repelled cations. A reinvestigation of his fixed-charge hypothesis in the sixties showed that the assumption of an accumulation of anions at the membrane surface by clusters of fixed positive charges seemed tenable, but that the fixed charges themselves could not act as a rate-limiting barrier as in *Mond's* original model. The nature of the barrier remained to be established (*Passow* 1969). After the discovery of the first pieces of evidence for a transport function of the band 3 protein in Toronto, Boston, and Frankfurt, *Rothstein* et al. (1976) suggested anion binding to a rotating side chain in an aqueous channel formed by the protein molecule. *Gunn* (1978) proposed a mechanism operating like a sluice with two gates in series. The anion enters the sluice with one gate open, the other closed. The open state closes, the anion passes by a positively charged group in the water-filled space between the two gates. The second gate opens, the anion is released at the other surface, and the sluice is capable of accepting another anion for the transfer in the opposite direction.

On the basis of the information reviewed above, it would not seem unreasonable to assume that several more or less parallel segments of the peptide chain form a channel that extends all the way across the membrane. The fluctuating energy barriers that constitute the rate-limiting steps are assumed to be confined to a narrow portion of the channel near the outer surface, where amino acid residues from the different adjacent segments of the peptide chain would form a gate. Some of the gate's constituents would also be constituents of the stilbene disulfonate binding site. In view of the high turnover number (about 5×10^4 s^{-1} band 3^{-1} for Cl^- at 38°C; *Brahm* 1977), it would seem likely that the rate-limiting steps are confined to a small number of fluctuating barriers. The fluctuations

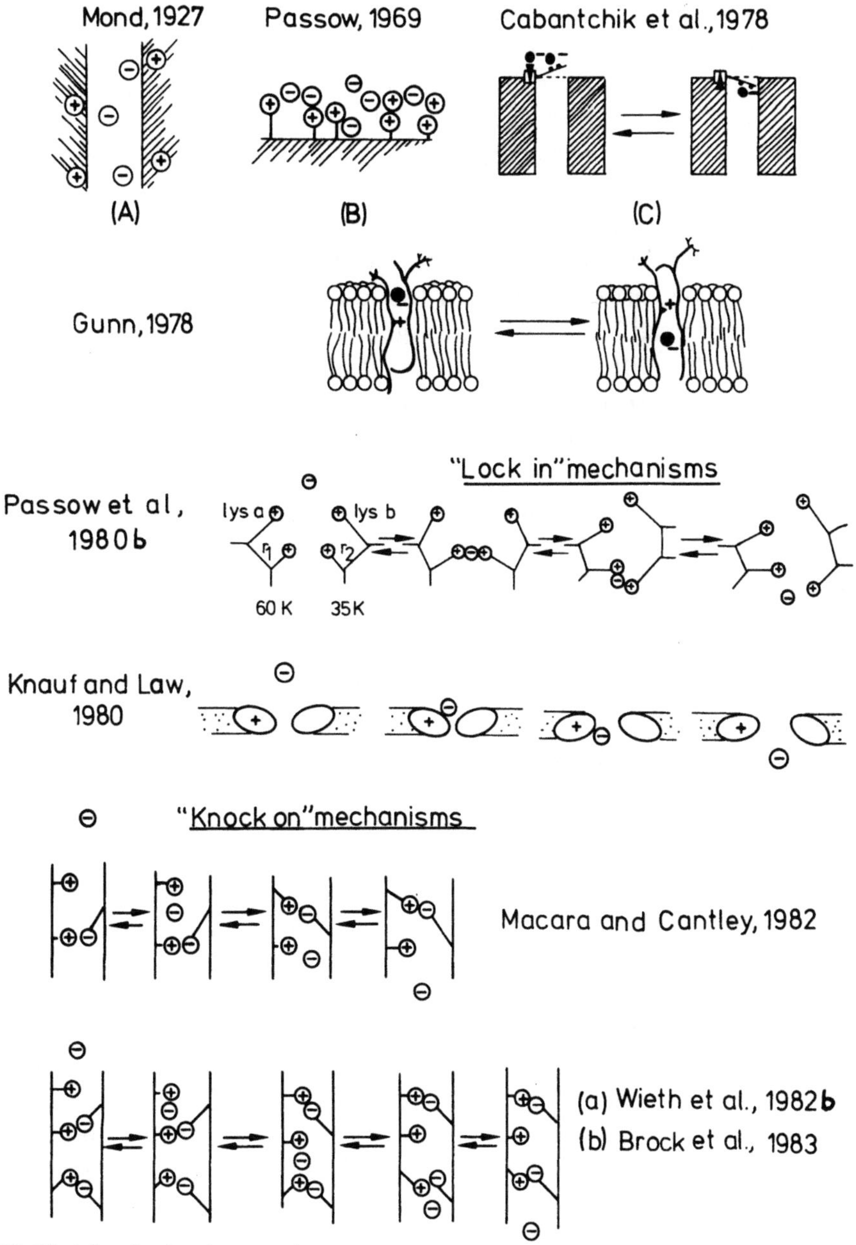

Fig. 29. Models of anion transport

would represent the allosterical and conformational changes during anion binding and translocation. The changes are not confined to the gate itself but can be monitored at distant regions of the molecule that are allosterically linked to the gate ("transfer site").

Currently there exist two essentially different types of models that may be designated as "knock on" and "lock in" models. Both models include the assumption that the substrate anions may be accumulated at the channel orifice by a cluster of positive fixed charges.

The "knock on" mechanism is based on the assumption that the arriving anion breaks salt bridges of protein-bound arginines and carboxyl groups by electrostatic interactions. The arriving anion is first directed to the salt bridge by the presence of specific arginine residues. It switches place with the carboxyl group of the ion pair. This leads to the establishment of a new salt bridge with the arginine residue that served to neutralize electrically the diffusible anion in front of the original salt bridge (*Macara* and *Cantly* 1983). In this model, the arriving anion reduces the height of the energy barrier in its penetration pathway; the migration of the anion would, therefore, contribute to the electrical conductance of the membrane. To account for the electrically nearly "silent" anion exchange, a similar model has been proposed with two such salt bridges in series (*Wieth* et al. 1982b; *Brock* et al. 1983): after passage of the first bridge and the reconstitution of a new bridge with the exofacial arginine residue, the second salt bridge is opened and subsequently closed after an exchange of places between the diffusible anion and the carboxyl groups involved in salt bridge formation. This model resembles that of *Gunn* (1978) and would not contribute to the conductance when one gate is closed before the other is opened. Since the penetrating anion opens a series of salt bridges, this model has also been called a "zipper model" (*Wieth* et al. 1982b).

According to the "lock in" mechanism, the arriving anion becomes entrapped by the originally open gate near the mouth of the channel (*Passow* et al. 1980b; *Knauf* et al. 1980) possibly at a site with one or two arginine residues. This would increase the barrier for anion diffusion across the channel. Transport would, therefore, require a reduction of the height of the energy barrier in front of the penetrating anion and, for the maintenance of a low conductance, an increase of the barrier in the rear. This would be the equivalent to the assumption that the complex between anion and transport protein undergoes a change from inward-facing to outward-facing conformation. The anion can now be released at the inner surface and the gate would be ready to bind the anion for the journey in the opposite direction.

In both types of models, the penetrating anion would catalyze the conformational transitions between inward-facing and outward-facing conformers. The small diffusive component would be explained by the "knock on" model, by the rare occurrence of the simultaneous opening of both of the consecutive gates, by the "lock in" model, and by the jump of the ion out of the substrate protein complex before the gate undergoes the conformational transition that leads to the electrically silent anion exchange.

At the present time it is difficult to express a clear preference for one or the other of the two types of models. Both models would require that the arginine residues at the outward-facing mouth of the channel are electrically neutralized by diffusible counter ions. Since these arginines do not seem to participate in recruitment phenomena, they are unlikely to be directly involved in anion binding. Thus, anion binding would need to take place more deeply inside the channel orifice.

The "knock on" process has been invoked on the grounds that the existence of ion pairs inside a pore traversing a membrane of low dielectric constant is energetically more favorable than the presence of ionic groups that are not electrically balanced (*Wieth* et al. 1982b; *Brock* et al. 1983). This argument is certainly true and renders mechanisms of the "knock on" type quite attractive. There exist, however, to date, no experimental observations that would be directly supportive for one or the other version of the knock on mechanism. Even the demonstrated participation of a carboxyl group in anion transport (*Bjerrum* 1983) is of little help in this respect since it is likely that it is involved in the maintenance of the functional state of the channel or gate rather than directly participating in anion binding (see p. 163).

The "lock in" mechanism would require that positively charged or uncharged amino acid residues at the gate replace the hydration shell of the substrate anion, similar to the cyclic carrier peptides that are known to solubilize hydrophilic cations in lipid double layers. This replacement would constitute the "lock in" process, which is then followed by a transition of the complex between the substrate and the constituents of the gate across the rate-limiting barrier and the release of the substrate into the channel leading to the inner membrane surface. The free enthalpy changes involved in such a process do not need to be much different from those seen with diffusable peptide carriers. Local disturbances of electroneutrality would also represent no obstacle for the operation of such a mechanism. There is no need for strict electrical neutrality at any location within the membrane. It is sufficient if electrical neutrality is maintained over the thickness of the membrane and the adherent electrical double layers.

In conclusion of this section it may be pertinent to recall the energy transfer experiments of *Macara* et al. (1983) discussed on p. 165. After covalent reaction of the nonpenetrating eosine maleimide with an amino acid residue on band 3 that is accessible from the outer surface, the eosine fluorescence can only be quenched by intracellular but not extracellular Cs^+. As has already been pointed out, this observation is difficult to reconcile with a Zipper model with several salt bridges in series, although it is easily compatible with a lock in mechanism.

Even after the amino acid sequence of band 3 has become known, it is not yet possible to replace the models described above by more realistic explanations of band 3 function in terms of band 3 structure. One of the main difficulties involved is the absence of adequate information about the three-dimensional structure of the band 3 protein and about the number, nature, and localization of the amino acid residues involved in anion binding and translocation. Nevertheless, the localization of certain covalently binding inhibitors of anion transport is known and can be used to make suggestions about the involvement of specific segments of the peptide chain in the transport process.

Lysine residue *a* in human band 3, which is allosterically linked to the substrate binding site and involved in the covalent bond formation with N_2ph-F and one of the isothiocyanate groups of H_2DIDS, is probably homologous to Lys-558 or Lys-561 in transmembrane segment 5 of mouse band 3 (*Kopito* and *Lodish* 1985). It is accessible from the external medium and allosterically linked to Lys-608, which resides in transmembrane segment 6 inside the bilayer and near the inner membrane surface (Fig. 26). Lysine residue *b*, which is also allosterically linked to the substrate binding site (reductive methylation leads to inhibition, *Jennings* and *Adams* 1981) and involved in the intramolecular cross-linking by H_2DIDS is located outside the segment sequenced by *Brock* et al. (1983). According to *Kopito* and *Lodish* (1985) Lys-657 is a likely candidate. It resides at the outer membrane surface in transmembrane segment 7. The involvement in anion transport of the loop formed by transmembrane segments 6 and 7 is further suggested by the work of *Jennings* et al. (1984) who have demonstrated a modification of anion transport after splitting the peptide chain with extracellular papain at Thr-647 in transmembrane segment 7. When H_2DIDS is added after splitting by papain, the intramolecular cross-link is no longer established between Lys *a* and Lys *b*, but between Lys *a* and some other lysine residue which in contrast to Lys *b* is situated inside the papain fragment P_7 (possibly Lys-643). This indicates that hydrolysis at Gly-648 leads to a rearrangement of a portion of the peptide chain that is associated with transport.

The results described make it appear likely that the putative transmembrane segments 5, 6, and 7 are constituents of an anion channel across the membrane.

Lys-449 at the outward-facing end of the transmembrane segment 1 can be modified without inhibition. This is probably also true for Cys-861 and -903 (although a clear demonstration of the absence of inhibition after a chemically verified success of the modification by SH reagents is still missing). Thus there is no evidence that the peptide chain between Arg-800 and the C-terminal Val-929 is involved in anion transport.

The arginine residues stipulated to be constituents of the stilbene disulfonate binding site and hence of the anion gate are possibly contributed by outward facing portions of segments 2, 4, 8, 9, or 10. These segments carry the arginine residues 451, 509, 675, 748, and 778 near the transitions from hydrophobic to hydrophilic regions of the peptide chain. Two of them could be involved in neutralizing the negative charges of the stilbene disulfonates and one or two more could account for the kinetics of the cross-linking by H_2DIDS of Lys-558 and either Lys-649 or -657. Although these possibilities constitute a suggestive basis for the formulation of hypotheses about the spatial arrangement of the peptide chain at the anion gate, the information at hand does evidently not suffice to make realistic suggestions. Thus, for the time being the knowledge about band 3 structure and the models derived from kinetics of transport cannot yet be bridged. The elucidation of the primary structure opens up, however, many new avenues for the establishment of a more realistic picture of the structural basis of the transport process.

Acknowledgements. I wish to thank Dr. D. Schubert, Dr. W. Schwarz, Dr. V. Rudloff, Dr. R. Peters, A. Berghout, and M. Raida for comments and criticism; Prof. H. Fasold for discussions about band 3 structure; Ms. S. Lepke and Ms. B. Legrum for their support during the preparation of the manuscript; and Mss. H. Zecher, G. Assar, S. Müller, and C. von Albrecht for their unfailing secretarial help during the various stages of this work. I also wish to express my gratitude to Drs. F. Sauer and K. Schnell who helped to clear my mind on many aspects of transport kinetics and to Dr. P.A. Knauf who provided extensive advice about the interpretation of certain experimental observations. Dr. B. Deuticke was kind enough to review the completed manuscript. I am very much indebted to Drs. R. Kopito and H. Lodish for their permission to quote from their work on the primary structure of band 3 prior to publication. The discussion of band 3 structure on pp. 75–79, 182–183 was written jointly with M. Raida.

The paper is dedicated to the memory of Jens Otto Wieth whose contributions have left lasting marks in the field reviewed above.

Appendix A

In the steady state the concentrations of all participants of the transport process are time-independent. For Ping-Pong kinetics without allosteric interactions between transfer site and modifier sites, this means that each of the derivatives with respect to time of *r, w, ar,* and *as* are zero. If, to be more general, slippage is included in the treatment (i.e., k_{18} and k_{81} are finite), one obtains:

$$\frac{dr}{dt} = -\alpha_1 - \beta_1 - k_{18}r + k_{81}s = 0 \tag{A1}$$

$$\frac{dar}{dt} = +\alpha_1 - k_{12}ar + k_{21}as = 0 \tag{A2}$$

$$\frac{dbr}{dt} = +\beta_1 - k_{15}br + k_{51}bs = 0 \tag{A3}$$

$$\frac{ds}{dt} = +\alpha_2 + \beta_2 + k_{18}r - k_{81}s = 0 \tag{A4}$$

$$\frac{das}{dt} = -\alpha_2 + k_{12}ar - k_{21}as = 0 \tag{A5}$$

$$\frac{dbs}{dt} = -\beta_2 + k_{15}br - k_{51}as = 0 \tag{A6}$$

The terms α_1, β_1 and α_2, β_2 account for the fact that due to the rapid establishment of the mass law equilibria ($a \cdot r = K_{101} \cdot ar$, $a \cdot s = K_{11} \cdot as$; $b \cdot r = K_{301} \cdot br$, $b \cdot s = K_{31} \cdot bs$), the s conformers that are converted into r conformers immediately react with a and b to form ar and br while equivalent amounts of ar and br that have been derived from as and bs are converted into $r + a + b$. Corresponding behavior applies to a, as, and bs. Adding equations A1, A2, and A3, or A4, A5, and A6 one obtains:

$$+\frac{d(r + ar + br)}{dt} = -\frac{d(s + as + bs)}{dt} = -k_{18}r + k_{81}s - k_{12}ar + k_{21}as - k_{15}br + k_{51}bs = 0$$

which upon rearrangement leads to:

$$k_{12}ar + k_{15}br + k_{18}r = k_{21}as + k_{51}bs + k_{81}s$$

Assuming that there is no slippage ($k_{18} = k_{81} = 0$) one arrives at the assumption (ii) (steady state) used in the derivation on p. 102.

The same type of reasoning also applies to all extensions of Ping-Pong kinetics involving modifier sites, e.g., Eqs. 2, 3, regardless of whether or not $a' = a'' = a$ and $b' = b'' = b$, or $a' \neq a''$ and $b' \neq b''$. It ensures that the requirement for microscopic reversibility is met.

Appendix B

The numerical evaluation of the equations of transport kinetics yields unique results only if the parameter values are independent of one another. To ensure this, for numerical calculations, it is necessary to reduce to the minimum the number of parameters in the equations that are required to describe the model. The usual algorithms for the solution of kinetic equations take this automatically into account. However, in the derivations given in this paper, where merging individual parameter values into composite quantities is deliberately avoided, the necessary reduction of the number of parameters has to be achieved by establishing all existing interrelationships, and subsequent elimination of parameters that can be defined in terms of other parameters.

The procedure may again be exemplified on the basis of Ping-Pong kinetics without modifier sites. The simplest possible reaction diagram including "slippage," i.e., the transitions $r \rightleftharpoons s$ of the unoccupied transfer site:

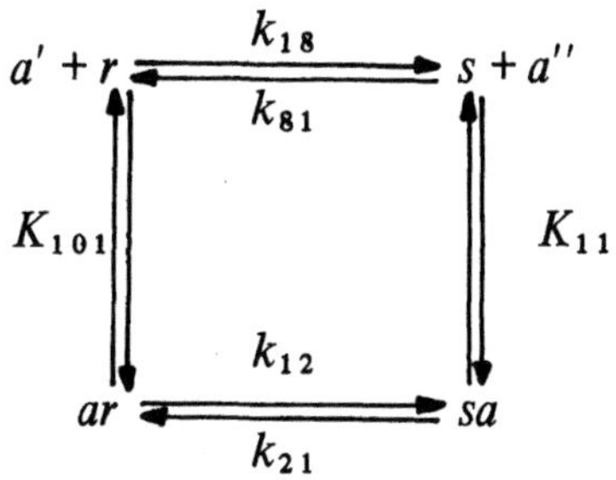

leads to the following expressions: $a \cdot r = K_{101} \cdot ar$, $a \cdot s = K_{11} \cdot as$, $r/s = k_{81}/k_{18}$, and $ar/as = k_{21}/k_{12}$. Combination of these equations yields: $K_{101} \cdot k_{21}/k_{12} = K_{11} \cdot k_{81}/k_{18}$. This enables one to eliminate one of the parameters, e.g., k_{81}/k_{18} by K_{101} k_{21}/k_{12} K_{11}.[13] For more complex systems, sets of similar relationships can be established and from their nature and number, the number of independently defined parameter values can be easily predicted. It is preferable not to execute all possible substitutions of excess parameters in the equations for j_{12} since this would tend to introduce terms that are not immediately intelligible, but to list them independently. This ensures that the relationship between the independent variable a and the various parameters that define the model as well as the relationships among the various parameters remain transparent and susceptible to the manipulations that are necessary for the exploration of the properties of the model.

It may be instructive to apply the reasoning described above to the most general reaction network used in this paper (Fig. 9b). This network is described by 24 mass law constants and 4 rate constants. The ratios k_{12}/k_{21} and k_{15}/k_{51} of the latter are equivalent to two more mass law constants. Thus there are altogether 26 constants which are related to each other by nine equations:

$$K_{400} \cdot K_{301} = K_{200} \cdot K_{401} \qquad K_{40} \cdot K_{31} = K_{20} \cdot K_{11}$$
$$K_{100} \cdot K_{102} = K_{101} \cdot K_{103} \qquad K_{10} \cdot K_{12} = K_{11} \cdot K_{13}$$
$$K_{300} \cdot K_{100} = K_{301} \cdot K_{303} \qquad K_{30} \cdot K_{10} = K_{31} \cdot K_{33}$$
$$K_{200} \cdot K_{202} = K_{101} \cdot K_{201} \qquad K_{20} \cdot K_{22} = K_{11} \cdot K_{21}$$

13 It should be noted that the rate constant k_{12} in the equation for the flux $j_{12} = k_{12} \cdot ar$ cannot be altered independent of k_{21}. For a given set of constants K_{101}, K_{11}, k_{18}/k_{81}, the ratio k_{21}/k_{12} is fixed. Hence a change of k_{12} requires a corresponding change of k_{21}

and

$$\frac{K_{101} \cdot K_{31}}{K_{11} \cdot K_{301}} = \frac{k_{12}}{k_{21}} \cdot \frac{k_{51}}{k_{15}}$$

These equations enable one to eliminate nine constants. The equation for j_{12} derived from the network (Eq. 3, p. 101) contains, in addition to k_{12}, ten constants. Thus, for numerical evaluation, one more constant needs to be eliminated. In the specific case at hand, it would be most convenient to eliminate either the ratio k_{12}/k_{15} or k_{21}/k_{51}.

Evidently, the parameter values used in Eq. 3 could be replaced by any combination of other parameter values in accordance with the relations listed above.

References

Acevedo F, Lundahl P. Fröman G (1981) The steriospecific D-glucose transport protein in cholate extracts of human erythrocyte membranes. Biochim Biophys Acta 648:254–262

Andersen OS, Bjerrum PJ, Borders CL, Broda T, Wieth JO (1983) Essential carboxyl groups in the anion exchange protein of human red blood cell membranes. Biophys J 41:164a

Appell KC, Low PS (1981) Partial structural characterization of the cytoplasmic domain of the erythrocyte membrane protein, band 3. J Biol Chem 256:11104–11111

Appell KC, Low PS (1982) Evaluation of structural independence of membrane-spanning and cytoplasmic domains of band 3. Biochemistry 21:2151–2157

Baldwin SA, Baldwin JM, Gorga FR, Lienhard GE (1979) Purification of the cytochalasin B binding component of the human erythrocyte monosaccharide transport system. Biochim Biophys Acta 552:183–188

Barzilay M, Cabantchik ZI (1979a) Anion transport in red blood cells II. Kinetics of reversible inhibition by nitroaromatic sulfonic acids. Membr Biochem 2:255–281

Barzilay M, Cabantchik ZI (1979b) Anion transport in red blood cells III. Sites and sidedness of inhibition by high affinity reversibly binding probes. Membr Biochem 2:297–322

Barzilay M, Ship S, Cabantchik ZI (1979) Anion transport in red blood cells I. Chemical properties of anion recognition sites as revealed by structure-activity relationships of aromatic sulfonic acids. Membr Biochem 2:227–254

Beigel M, Loyter A (1983) Fusion-mediated implantation of band 3 into living cells. A new system to study degradation of membrane protein. Exp Cell Res 148:95–103

Bennett V, Stenbuck PJ (1979) The membrane attachment protein for spectrin is associated with band 3 in human erythrocyte membranes. Nature 280:468–473

Bennett V, Stenbuck PJ (1980) Association between ankyrin and the cytoplasmic domain of band 3 isolated from the human erythrocyte membrane. J Biol Chem 255:6424–6432

Bentley PJ, McGahan MC (1980) Inhibitory action of DIDS on chloride transport across the amphibian cornea. J Physiol 304:519–527

Bentley PJ, McGahan MC (1982) A pharmacological analysis of chloride transport across the amphibian cornea. J Physiol 325:481–492

Benz R, Tosteson MT, Schubert D (1984) Formation and properties of tetramers of band 3 protein from human erythrocyte membranes in planar lipid bilayers. Biochem Biophys Acta 775:347–355

Berghout A, Raida M, Romano L, Passow H (1984) Inverse effects of dansylation of the red blood cell on band 3-mediated transport of monovalent and divalent anions. Hoppe Seyler's Z Physiol Chem 365:226

Berghout A, Raida M, Romano L, Passow H (1985) pH dependence of phosphate transport across the red blood cell membrane after modification by dansyl chloride. Biochim Biophys Acta 815:281–286

Bittar EE, Schultz R, Tesar J (1980) Chloride efflux in single barnacle muscle fibres. J Physiol 301:317–336

Bjerrum P (1983) Identification and location of amino acid residues essential for anion transport in red cell membranes. In: Quagliariello E, Palmieri F (eds) Structure and function of membrane proteins. Elsevier, Amsterdam, pp 107–115

Bjerrum PJ, Tranum-Jensen J, Møllgård (1980) Morphology of erythrocyte membranes and their transport functions following aggregation of membrane proteins. In: Lassen UV, Ussing HH, Wieth JO (eds) Membrane transport in erythrocytes. Munksgaard, Copenhagen, pp 51–68 (Alfred Benzon Symposium 14)

Bjerrum PJ, Wieth JO, Borders CL (1983) Selective phenylglyoxalation of functionally essential arginyl residues in the erythrocyte anion transport protein. J Gen Physiol 81:453–484

Boodhoo A, Reithmeier RAF (1983) Characterization of matrix-bound band 3, the anion transport protein from human erythrocyte membranes. J Biol Chem 259: 785–790

Borochov H, Shinitzky M (1976) Vertical displacement of membrane proteins mediated by changes in microviscosity. Proc Natl Acad Sci USA 73:4526–4530

Borochov H, Abbott RE, Schachter D, Shinitzky M (1976) Modulation of erythrocyte membrane proteins by membrane cholesterol and lipid fluidity. Biochemistry 18: 251–255

Braell WA, Lodish HF (1981) Biosynthesis of the erythrocyte anion transport protein. J Biol Chem 256:11337–11344

Braell WA, Lodish HF (1982) The erythrocyte anion transport protein is cotranslationally inserted into microsomes. Cell 28:23–31

Brahm J (1977) Temperature-dependent changes of chloride transport kinetics in human red blood cells. J Gen Physiol 70:283–306

Brazy PC, Gunn RB (1976) Furosemide inhibition of chloride transport in human red blood cells. J Gen Physiol 68:583–599

Brock CJ, Tanner MJA, Kempf C (1983) The human erythrocyte anion-transport protein. Biochem J 213:577–586

Brodsky WA, Durham J, Ehrenspeck G (1979) The effects of a disulphonic stilbene on chloride and bicarbonate transport in the turtle bladder. J Physiol 287:559–573

Brown CDA, Simmons NL (1981) Catecholamine-stimulation of Cl^--secretion in MDCK cell epithelium. Biochim Biophys Acta 649:427–435

Brown PA, Feinstein MB, Sha'afi RI (1975) Membrane proteins related to water transport in human erythrocytes. Nature 254:523–525

Cabantchik ZI, Rothstein A (1974a) Membrane proteins related to anion permeability of human red blood cells I. Localisation of disulfonic stilbene binding sites in proteins involved in permeation. J Membr Biol 15:207–226

Cabantchik ZI, Rothstein A (1974b) Membrane proteins related to anion permeability of human red blood cells II. Effects of proteolytic enzymes on disulfonic stilbene sites of surface proteins. J Membr Biol 15:227–248

Cabantchik ZI, Balshin M, Breuer W, Rothstein A (1975) Pyridoxal phosphate. An anionic probe for protein amino groups exposed on the outer and inner surfaces of intact human red blood cells. J Biol Chem 250:5130–5136

Cabantchik ZI, Knauf PA, Rothstein A (1978) The anion transport system of the red blood cell. The role of membrane protein evaluated by use of "probes". Biochim Biophys Acta 515:239–302
Cabantchik ZI, Volsky DJ, Ginsburg H, Loyter A (1980) Reconstitution of the erythrocyte anion transport system. Ann NY Acad Soc 341:444–454
Canfield VA, Macey RI (1984) Anion exchange in human erythrocytes has a large activation volume. Biochim Biophys Acta 778:379–384
Carter-Su C, Pessin JE, Mora R, Gitomer W, Czech MP (1982) Photoaffinity labeling of the human erythrocyte D-glucose transporter. J Biol Chem 257:5419–5425
Cassoly R (1983) Quantitative analysis of the association of human hemoglobin with the cytoplasmic fragment of band 3 protein. J Biol Chem 258:3859–3864
Cassoly R, Salhany JM (1983) Spectral and oxygen-release kinetic properties of human hemoglobin bound to the cytoplasmic fragment of band 3 protein in solution. Biochim Biophys Acta 745:134–139
Chasan B, Lukacovic MF, Toon MR, Solomon AK (1984) Effect of thiourea on pCMBS inhibition of osmotic water transport in human red cells. Biochim Biophys Acta 778:185–190
Cherry RJ, Nigg EA (1980) Molecular interactions involving band 3: Information from rotational diffusion measurements. In: Lassen UV, Ussing HH, Wieth JO (eds) Membrane transport in erythrocytes. Mungksgaard, Copenhagen, pp 130–138 (Alfred Benzon Symposium 14)
Cherry RJ, Bürkli A, Busslinger M, Schneider G, Parish GR (1976) Rotational diffusion of band 3 protein in the human red cell membrane. Nature 263:389–393
Clarke S (1975) The size and detergent binding of membrane proteins. J Biol Chem 250:5459–5469
Cleland WW (1963) The kinetics of enzyme-catalyzed reactions with two or more substrate or products, I. Nomenclature and rate equations. Biochim Biophys Acta 67: 104–137
Cousin JL, Motais R (1979) Inhibition of anion permeability by amphiphilic compounds in human red cells: evidence for interactions of niflumic acid with the band 3 protein. J Membr Biol 46:125–153
Cousin JL, Motais R (1982a) Inhibition of anion transport in the red blood cell by anionic amphiphilic compounds I. Determination of the flufenamate-binding site by proteolytic dissection of the band 3 protein. Biochim Biophys Acta 687:147–155
Cousin JL, Motais R (1982b) Inhibition of anion transport in the red blood cell by anionic amphiphilic compounds II. Chemical properties of the flufenamate-binding site on the band 3 protein. Biochim Biophys Acta 687:156–164
Cox JV, Moon RT, Lazarides E (1985) Anion transporter: highly cell-type-specific expression of distinct polypeptides and transcripts in erythroid and non-erythroid cells. J Cell Biol 100:1548–1557
Craik JD, Reithmeier RAF (1984) Inhibition of phosphate transport in human erythrocytes by water-soluble carbodiimides. Biochim Biophys Acta 778:429–434
Dalmark M (1975) Chloride transport in human red cells. J Physiol 250:39–64
Dalmark M (1976) Effects of halides and bicarbonate on chloride transport in human red blood cells. J Gen Physiol 67:223–234
Darmon A, Zangvill M, Cabantchik ZI (1983) New approaches for the reconstitution and functional assay of membrane transport proteins. Application to the anion transporter of human erythrocytes. Biochim Biophys Acta 727:77–88
Dekowski SA, Rybicki A, Drickamer K (1983) A tyrosine kinase associated with the red cell membrane phosphorylates band 3. J Biol Chem 258:2750–2753
Deuticke B (1977) Properties and structural basis of simple diffusion pathways in the erythrocyte membrane. Rev Physiol Biochem Pharmacol 78:1–97
Deuticke B, von Benthheim M, Beyer E, Kamp D (1978) Reversible inhibition of anion exchange in human erythrocytes by an inorganic disulfonate, tetrathionate. J Membr Biol 44:135–158

Deuticke B, Grunze M, Haest CWM (1980) Influence of membrane lipids on ion and non electrolyte transport through the erythrocyte membrane. In: Lassen UV, Ussing HH, Wieth JO (eds) Membrane transport in erythrocytes. Munksgaard, Copenhagen, pp 143–156 (Alfred Benzon Symposium 14)
Deziel M, Pegg W, Mack E, Rothstein A, Klip A (1984) Labelling of the human erythrocyte glucose transporter with ^{3}H-labelled cytochalasin B occurs via protein photoactivation. Biochim Biophys Acta 772:403–406
Dissing S, Romano L, Passow H (1981) The kinetics of anion equilibrium exchange across the red blood cell membrane as measured by means of ^{35}S-thiocyanate. J Membr Biol 62:219–229
Dix JA, Verkman AS (1982) Stopped-flow and temperature-jump measurements on biological systems: effects of heterogeneity, unstirred layers and multiple reaction intermediates. Biophys J 37:216a
Dix JA, Verkman AS, Solomon AK, Cantley LC (1979) Human erythrocyte anion exchange site characterized using a fluorescent probe. Nature 282:520–522
Dix JA, Verkman AS, Solomon AK (1981) "Rate constant inhibition" – a possible mechanism for stilbene inhibition of anion exchange. Biophys J 33:49a
Dorst H-J, Schubert D (1979) Self-association of band 3 protein from human erythrocyte membranes in aqueous solutions. Hoppe-Seyler's Z Physiol Chem 360:1605–1618
Drickamer LK (1976) Fragmentation of the 95,000-Dalton transmembrane polypeptide in human erythrocyte membranes. J Biol Chem 251:5115–5123
Drickamer LK (1977) Fragmentation of the band 3 polypeptide from human erythrocyte membranes. J Biol Chem 252:6906–6917
Drickamer LK (1978) Orientation of the band 3 polypeptide from human erythrocyte membranes. J Biol Chem 253:7242–7248
Du Pre AM, Rothstein A (1981) Inhibition of anion transport associated with chymotryptic cleavages of red blood cell band 3 protein. Biochim Biophys Acta 646:471–478
Edidin M (1981) Molecular motions and membrane organization and function. In: Finean FB, Michell RH (eds) Membrane structure. Elsevier/North Holland, Amsterdam, pp 37–82 (New comprehensive biochemistry, vol 1)
Ehrenspeck G (1982) Effect of 3-isobutyl-1-methylxanthine on HCO_3-transport in turtle bladder. Biochim Biophys Acta 684:219–227
Eidelman D, Cabantchik ZI (1983a) The mechanism of anion transport across human red blood cell membranes as revealed with a fluorescent substrate. II Kinetic properties of NBD-taurine transfer in symetric conditions. J Membr Biol 71:141–148
Eidelman D, Cabantchik ZI (1983b) The mechanism of anion transport across human red blood cell membranes as revealed with a fluorescent substrate II kinetic properties of NBD-taurine transfer in asymmetric conditions. J Membr Biol 71:149–161
Eisenberg D (1984) Three-dimensional structure of membrane and surface proteins. Ann Rev Biochem 53:595–623
Eisenberg D, Schwarz E, Komaromy M, Wall R (1984) Analysis of membrane and surface protein sequences with the hydrophobic moment plot. J Mol Biol 179:125–142
Eisinger J, Flores J, Salhany JM (1982) Association of cytosol hemoglobin with the membrane in intact erythrocytes. Proc Natl Acad Sci 79:408–412
Falke FJ, Chan SI (1984) Ion channels within ion transport proteins. Evidence in the band 3 system. Biophys J 45:91–92
Falke JF, Pace RJ, Chan SI (1984a) Chloride binding to the anion transport binding sites of band 3. A ^{35}Cl NMR study. J Biol Chem 259:6472–6480
Falke JF, Pace RJ, Chan SI (1984b) Direct observation of the transmembrane recruitment of band 3 transport sites by competitive inhibitors. A ^{35}Cl NMR study. J Biol Chem 259:6481–6491
Findlay JBC (1974) The receptor proteins for concanavalin A and lens culinaris phytohemmagglutinin in the membrane of the human erythrocyte. J Biol Chem 249: 4398–4403

Formann SA, Verkman AS, Dix JA, Solomon AK (1981) Phloretin binds to band 3, the anion transport protein of the red blood cell membrane. Biophys J 33:48a
Formann SA, Verkman AS, Dix JA, Solomon AK (1982) Effect of lipid perturbants on red cell band 3 conformational states. Biophys J 37:216a
Fröhlich O (1982) The external anion binding site of the human erythrocyte anion transport: DNDS binding and competition with chloride. J Membr Biol 65:111–123
Fröhlich O (1983) Contributions of slippage and tunneling to anion net transport across the human red blood cell membrane. Biophys J 41:63a
Fröhlich O (1984a) How channel-like is a biological carrier. Studies with the erythrocyte anion transporter. Biophys J 45:93–94
Fröhlich O (1984b) Relative contributions of the slippage and tunneling mechanisms to anion net efflux from human erythrocytes. J Gen Physiol 84:877–893
Fröhlich O, Gunn RB (1982) Mutual interactions of reversible inhibitors on the red cell anion transporter. Biophys J 37:213a
Fröhlich O, Leibson C, Gunn RB (1983) Evidence for a positive charge on the anion binding/transport site. J Gen Physiol 81:127–152
Fukuda M, Eshdat Y, Tarone G, Marchesi VT (1978) Isolation and characterization of peptides derived from the cytoplasmic segment and band 3, the predominant intrinsic membrane protein of the human erythrocyte. J Biol Chem 253:2419–2428
Fukuda MN, Fukuda M, Hakamori S (1979) Cell surface modification by endo-β-galactosidase. J Biol Chem 254:5458–5465
Fukuda M, Dell A, Oates JE, Fukuda MN (1984) Structure of branched lactoseaminoglycan, the carbohydrate moiety of band 3 isolated from adult human erythrocytes. J Biol Chem 259:8260–8273
Furuya Y, Tarshis T, Law F-Y, Knauf PA (1984) Transmembrane effects of intracellular chloride on the inhibitory potency of extracellular H_2DIDS. Evidence for two conformations of the transport site of the human erythrocyte anion exchange protein. J Gen Physiol 83:657–681
Funder J, Wieth JO (1976) Chloride transport in human erythrocytes and ghosts: a quantitative comparison. J Physiol 262:679–698
Galvez LM, Jennings M, Tosteson M (1984) Incorporation of the DIDS binding peptide from the anion transport protein into bilayers. Fed Proc 43:315
Geck P, Pietrzyk C, Burckhardt B-C, Pfeiffer B, Heinz E (1980) Electrically silent cotransport of Na^+, K^+ and Cl^- in Ehrlich cells. Biochim Biophys Acta 600:432–447
Golan DE, Veatch W (1980) Lateral mobility of band 3 in the human erythrocyte membrane studied by fluorescence photobleaching recovery: evidence for control by cytoskeletal interactions. Proc Natl Acad Sci 77:2537–2541
Golan DE, Veatch W (1982) Lateral mobility of band 3 in the human erythrocyte membrane: control by ankyrin-mediated interactions. Biophys J 37:177a
Grinstein S, Ship S, Rothstein A (1978) Anion transport in relation to proteolytic dissection of band 3 protein. Biochim Biophys Acta 507:294–304
Grinstein S, McCullough L, Rothstein A (1979) Transmembrane effects of irreversible inhibitors of anion transport in red blood cells. Evidence for mobile transport sites. J Gen Physiol 73:493–514
Gruber W, Deuticke B (1973) Comparative aspects of phosphate transfer across mammalian erythrocyte membranes. J Membr Biol 13:19–36
Grunze M, Forst B, Deuticke B (1980) Dual effect of membrane cholesterol on simple and mediated transport processes in human erythrocytes. Biochim Biophys Acta 600:860–869
Grygorczyk R, Schwarz W (1983) Properties of the Ca^{2+}-activated K^+ conductance of human red cells, as revealed by the patch clamp technique. Cell Calcium 4:499–510
Grygorczyk R, Schwarz W, Passow H (1984) Ca^{++}-induced K^+ channels in human red cells: a comparison of single channel currents with ion fluxes. Biophys J 45:693–698

Guidotti G (1977) The structure of intrinsic membrane proteins. J Supermol Struct 7:489–497

Guidotti G (1980) The structure of the band 3 polypeptide. In: Lassen UV, Ussing HH. Wieth JO (eds) Membrane transport in erythrocytes. Munksgaard, Copenhagen, pp 300–308 (Alfred Benzon Symposium 14)

Gunn RB (1973) A titratable carrier model for monovalent and divalent inorganic anions in red blood cells. In: Gerlach E, Moser K, Deutsch E, Wilmanns W (eds) Erythrocytes, thrombocytes, leucocytes. Thieme, Stuttgart, pp 77–79

Gunn RB (1978) Considerations of the titratable carrier model for sulfate transport in human red blood cells. In: Hoffman JF (ed) Membrane transport processes, vol 1. Raven, New York, pp 61–77

Gunn RB (1979) Transport of anions across red cell membranes. In: Giebisch G, Tosteson DC, Ussing HH (eds) Transport across biological membranes of membrane transport in biology, vol II. Springer, Berlin Heidelberg New York, pp 59–80

Gunn RB, Fröhlich O (1979) Assymetry in the mechanism for anion exchange in human red blood cell membranes. Evidence for reciprocating sites that react with one transported anion at a time. J Gen Physiol 74:351–374

Gunn RB, Fröhlich O (1982) Arguments in support of a single transport site on each anion transporter in human red cells. In: Zadunaisky J (ed) Chloride transport in biological membranes. Academic, New York, pp 33–59

Gunn RB, Milanick MA (1982) Internal protons are competitive inhibitors of chloride exchange in human erythrocytes. Biophys J 37:336a

Gunn RB, Milanick MA (1983) Internal protons are a mixed inhibitor of chloride exchange in human erythrocytes. Fed Proc 42:606

Gunn RB, Dalmark M, Tosteson D, Wieth JO (1973) Characteristics of chloride transport in human red blood cells. J Gen Physiol 61:185–206

Gunn RB, Fröhlich O, Macintyre JD, Low PS (1979) Calcium modification of the anion transport mechanism in red blood cells. Biophys J 25:106a

Haest CWM (1982) Interactions between membrane skeleton proteins and the intrinsic domain of the erythrocyte membrane. Biochim Biophys Acta 694:331–352

Haest CWM, Kamp D, Plasa G, Deuticke B (1977) Intra- and intermolecular cross-linking of membrane proteins in intact erythrocytes and ghosts by SH-oxidizing agents. Biochim Biophys Acta 469:226–230

Halestrap AP (1976) Transport of pyruvate and lactate in human erythrocytes. Evidence for the involvement of the chloride carrier and a chloride-independent-carrier. Biochem J 156:193–207

Hargreaves WR, Giedd KN, Verkleji A, Branton D (1980) Reassociation of ankyrin with band 3 in erythrocyte membranes and in lipid vesicles. J Biol Chem 255: 11965–11972

Harriw EJ, Pressman BC (1967) Obligate cation exchanges in red cells. Nature 216: 918–920

Hautmann M, Schnell KF (1985) Concentration dependence of the chloride self exchange and homoexchange fluxes in human red cell ghosts. Pflügers Arch 405:193–201

Herbst F, Rudloff V (1982) Acylation of integral erythrocytes membrane proteins resulting in a soluble form of band 3 protein. Protides Biol Fluids 29:113–116

Herbst F, Rudloff V (1984) The preparation of hydrophilic derivatives of band 3 protein by acylation of the human red blood cell membrane. Hoppe-Seyler's Z Physiol Chem 365:525–530

Higashi T, Richards CS, Uyeda K (1979) The interaction of phosphofructokinase with erythrocyte membranes. J Biol Chem 254:9542–9550

Ho MK, Guidotti G (1975) A membrane protein from human erythrocytes involved in anion exchange. J Biol Chem 250:675–685

Hoffman JF, Laris PC (1974) Determination of membrane potentials in human and amphibian red blood cells by means of a fluorescent probe. J Physiol 239:519–552

Hoffman JF, Laris PC (1984) Membrane electrical parameters of normal human red blood cells. In: Blaustein MP, Lieberman M (eds) Electrogenic transport: fundamental principles and physiological implications. Raven, New York, pp 287–293
Hoffman JF, Lassen UV (1970) Plasma membrane potentials in amphiuma red cells. Proc XXV Congr Int Union Physiol Sci IX:253
Hoffman JF, Kaplan JH, Callahan TJ, Freedman JC (1980) Electrical resistance of the red cell membrane and the relation between net anion transport and the anion exchange mechanism. Ann NY Acad Sci 341:357–360
Hsu L, Morrison M (1983) The interaction of human erythrocyte band 3 with cytoskeletal components. Arch Biochem Biophys 227:31–38
Hunter MJ (1971) A quantitative estimate of the non-exchange-restricted chloride permeability of the human red cell. J Physiol 218:49–50
Hunter MJ (1977) Human erythrocyte anion permeabilities measured under conditions of net charge transfer. J Physiol 268:35–49
Jackson P, Morgan B (1982) The relation between the membrane cholesterol content and anion exchange in the erythrocytes of patients with cholistasis. Biochim Biophys Acta 693:99–104
Jacquez JA (1964) The kinetics of carrier-mediated transport. Biochim Biophys Acta 79:318–328
Jenkins RE, Tanner MJA (1977) The structure of the major protein of the human erythrocyte membrane. Characterization of the intact protein and major fragments. Biochem J 161:139–147
Jennings ML (1976) Proton fluxes associated with erythrocyte membrane anion exchange. J Membr Biol 28:187–205
Jennings ML (1978) Characteristics of CO_2-independent pH equilibration in human red blood cells. J Membr Biol 40:365–391
Jennings ML (1980) Apparent "recruitment" of SO_4 transport sites by the Cl gradient across the human erythrocyte membrane. In: Lassen UV, Ussing HH, Wieth JO (eds) Membrane transport in erythrocytes. Munksgaard, Copenhagen, pp 450–463 (Alfred Benzon Symposium 14)
Jennings ML (1982a) Reductive methylation of the two H_2DIDS-binding lysine residues of band 3, the human erythrocyte anion transport protein. J Biol Chem 257:7554–7559
Jennings ML (1982b) Stoichiometry of a half-turnover of band 3, the chloride transport protein of human erythrocytes. J Gen Physiol 79:169–185
Jennings ML (1984) Oligomeric structure and the anion transport function of human erythrocyte band 3 protein. J Membr Biol 80:105–117
Jennings ML (1985) Kinetics and mechanism of anion transport in red blood cells. Annu Rev Physiology 47:519–533
Jennings ML, Adams MF (1981) Modification by papain of the structure and function of band 3, the erythrocyte anion transport protein. Biochemistry 20:7118–7122
Jennings ML, Nicknish JS (1984) Erythrocyte band 3 protein: evidence for multiple membrane-crossing segments in the 17000-dalton chymotryptic fragment. Biochemistry 23:6432–6436
Jennings ML, Passow H (1979) Anion transport across the erythrocyte membrane, in situ proteolysis of band 3 protein, and cross-linking of proteolytic fragments by 4,4′-diisothiocyano-dihydrostilbene-2,2′-disulfonate. Biochim Biophys Acta 554: 498–519
Jennings ML, Adams-Lackey M, Denney GH (1984) Peptides of human erythrocyte band 3 protein produced by extracellular papain cleavage. J Biol Chem 259:4652–4660
Kampman L, Lepke S, Fasold H, Fritzsch G, Passow H (1982) The kinetics of intramolecular cross-linking of the band 3 protein in the red blood cell membrane by 4,4′-diisothiocyano dihydrostilbene-2,2′-disulfonic acid (H_2DIDS). J Membr Biol 70:199–216

Kapitza H-G, Sackmann E (1980) Local measurement of lateral motion in erythrocyte membranes by photobleaching technique. Biochim Biophys Acta 595:56–64

Kaplan JH (1972) Anion diffusion across artificial lipid membranes: the effects of lysozyme on anion diffusion from phospholipid liposomes. Biochim Biophys Acta 290:339–347

Kaplan JH, Scorah K, Fasold H, Passow H (1976) Sidedness of the inhibitory action of disulfonic acids on chloride equilibrium exchange and net transport across the human erythrocyte membrane. FEBS Lett 62:182–185

Kaplan JH, Pring M, Passow H (1980) Concentration dependence of chloride movements that contribute to the conductance of the red cell membrane. In: Lassen UV, Ussing HH, Wieth JO (eds) Membrane transport in erythrocytes. Munksgaard, Copenhagen, pp 494–497 (Alfred Benzon Symposium 14)

Kaplan JH, Pring M, Passow H (1982) Band-3-mediated diffusive anion flow across the red blood cell membrane. Fed Proc 41:975

Kaplan JH, Pring M, Passow H (1983) Band-3 protein-mediated anion conductance of the red cell membrane. FEBS Lett 156:175–179

Karadsheh NS, Uyeda K (1977) Changes in alosteric properties of phosphofructokinase bound to erythrocyte membranes. J Biol Chem 252:7418–7420

Kasahara M, Hinkle PC (1977) Reconstitution and purification of the D-glucose transporter from human erythrocytes. J Biol Chem 252:7384–7390

Kaul RK, Murthy SNP, Reddy AG, Steck TL, Köhler H (1983) Amino acid sequence of the N-terminal 201 residues of human erythrocyte membrane band 3. J Biol Chem 258:7981–7990

Kay MMB, Tracey CM, Goodman JR, Cone JC, Bassel PS (1983) Polypeptides immunologically related to band 3 are present in nucleated somatic cells. Proc Natl Acad Sci USA 80:6882–6886

Kelly GE, Winzor DJ (1984) Quantitative characterization of the interactions of aldolase and glyceraldehyde-3-phosphate dehydrogenase with erythrocyte membranes. Biochim Biophys Acta 778:67–73

Kempf C, Brock C, Sigrist H, Tanner MJA, Zahler P (1981) Interaction of phenylisothiocyanate with human erythrocyte band 3 protein. II Topology of phenylisothiocyanate modification. Biochim Biophys Acta 641:88–98

Kiehm DJ, Ji TH (1977) Photochemical cross-linking of cell membranes. J Biol Chem 252:8524–8531

Kimelberg HK (1981) Active accumulation and exchange transport of chloride in astroglial cells in culture. Biochim Biophys Acta 646:179–184

Kleinmann JG, Ware RA, Schwartz JH (1981) Anion transport regulates intracellular pH in renal cortical tissue. Biochim Biophys Acta 648:87–92

Klimann HJ, Steck TL (1980a) Kinetic analysis of the association of glyceraldehyde 3-phosphate dehydrogenase with the human red cell membrane. In: Lassen UV, Ussing HH, Wieth JO (eds) Membrane transport in erythrocytes. Munksgaard, Copenhagen, pp 312–322 (Alfred Benson Symposium 14)

Klimann HJ, Steck TL (1980b) Association of glyceraldehyde-3-phosphate dehydrogenase with the human red cell membrane. A kinetic analysis. J Biol Chem 255: 6314–6321

Klugerman AH, Gaarn A, Parkes JG (1984) Effect of cholesterol upon the conformation of band 3 and its transmembrane fragment. Can J Biochem Cell Biol 62:1033–1040

Knauf PA (1979) Erythrocyte anion exchange and the band 3 protein; transport kinetics and molecular structure. Curr Top Membr Transp 12:249–363

Knauf PA (1981) Niflumic acid senses the conformation of the transport site of the human red cell anion exchange system. Biophys J 33:49a

Knauf PA (1982) Kinetic assymetry of the red cell anion exchange system. In: Martonosi A (ed) Membranes and transport, vol II. Plenum, New York, pp 441–449

Knauf PA (to be published) Anion transport in erythrocytes. In: Andreoli T, Hoffman JF, Schultz SG, Fanenstil DD (eds) Membrane transport disorders, 2nd Ed. Plenum, New York

Knauf PA, Law FY (1980) Relationship of net anion flow to the anion exchange system. In: Lassen UV, Ussing HH, Wieth JO (eds) Membrane transport in erythrocytes. Munksgaard, Copenhagen, pp 448–493 (Alfred Benson Symposium 14)

Knauf PA, Mann N (1982) Use of niflumic acid (NA) to probe the asymmetry of the human erythrocyte anion exchange systems. Fed Proc 41:975

Knauf PA, Mann NA (1984a) Location of the modifier site of the human erythroyte anion exchange system. Biophys J 45:18a

Knauf PA, Mann NA (1984b) Use of niflumic acid to determine the nature of the asymmetry of the human erythrocyte anion exchange system. J Gen Physiol 83: 703–725

Knauf PA, Rothstein A (1971) Chemical modification of membranes. In: Effects of sulfhydryl and amino reactive reagents on anion and cation permeability of the human red blood cell. J Gen Physiol 58:190–210

Knauf PA, Fuhrmann GF, Rothstein S, Rothstein A (1977) The relationship between anion exchange and net anion flow across the human red blood cell membrane. J Gen Physiol 69:363–386

Knauf PA, Breuer W, McCulloch L, Rothstein A (1978a) *N*-(4-azido-2-nitrophenyl)-2-aminoethylsulfonate (NAP-taurine) as a photoaffinity probe for identifying membrane somponents containing the modifier site of the human red blood cell anion exchange system. J Gen Physiol 72:631–649

Knauf PA, Ship S, Breuer W, McCulloch L, Rothstein A (1978b) Asymmetry of the red cell anion exchange system: different mechanism of reversible inhibition by *N*-(4-azido-2-nitrophenyl)-2-aminoethyl-sulfonate (NAP-taurine) at the inside and outside of the membrane. J Gen Physiol 72:607–630

Knauf PA, Tarshis T, Grinstein S, Furuya W (1980) Spontaneous and induced asymmetry of the human erythrocyte anion exchange system as detected by chemical probes. In: Lassen UV, Ussing HH, Wieth JO (eds) Membrane transport in erythrocytes. Munksgaard, Copenhagen, pp 389–403 (Alfred Benzon Symposium 14)

Knauf PA, Mann N, Law F-Y (1981) Niflumic acid senses the conformation of the transport site of the human red cell anion exchange system. Biophys J 33:49a

Knauf PA, Law F-Y, Marchant PJ (1983a) Relationship of net chloride flow across the human erythrocyte membrane to the anion exchange mechanism. J Gen Physiol 81:95–126

Knauf PA, Mann NA, Kalwas JE (1983b) Net chloride trânsport across the human erythrocyte membrane into low chloride media: evidence against a slippage mechanism. Biophys J 41:164a

Knauf PA, Law FY, Tarshis T, Furuya W (1984) Effects of the transport site conformation on the binding of external NAP-taurine to the human erythrocyte anion exchange system. J Gen Physiol 83:683–701

Köhne W, Haest CWM, Deuticke B (1981) Mediated transport of anions in band 3-phospholipid vesicles. Biochim Biophys Acta 229:547–556

Köhne W, Deuticke B, Haest CWM (1983) Phospholipid dependence of the anion transport system of the human erythrocyte membrane. Biochim Biophys Acta 730: 139–150

Kopito RR, Lodish HF (1985) Primary structure and transmembrane orientation of the murine anion exchange protein. Nature 316:234–238

Koppel DE, Sheetz MP, Schindler M (1981) Matrix control of protein diffusion in biological membranes. Proc Natl Acad Sci 78:3576–3580

Ku CP, Jennings ML, Passow H (1979) A comparison of the inhibitory potency of reversibly acting inhibitors of anion transport on chloride and sulfate movements across the human red cell membrane. Biochim Biophys Acta 553:132–144

Läuger P (1980) Kinetic properties of ion carriers and channels. J Membr Biol 57: 163–178

Läuger P (1984) Channels with multiple conformational states: interrelations with carriers and pumps. Curr Tob Membr Trans 21:309–326

Läuger P (1985) Ionic channels with conformational substates. Biophys J 47:581–590

Läuger P, Stephan W, Frehland E (1980) Fluctuations of barrier structure in ionic channels. Biochim Biophys Acta 602:167–180

Langridge-Smith JE, Field M (1981) Sulfate transport in rabbit ileum: characterization of the serosal border anion exchange process. J Membr Biol 63:207–214

Lassen UV (1972) Membrane potential and membrane resistance of red cells. In: Rørth M, Astrup P (eds) Oxygen affinity and red cell acid-base status. Munksgaard, Copenhagen, pp 291–304 (Alfred Benzon Symposium IV)

Lassen UV (1977) Electrical potential and conductance of the red cell membrane. In Ellory JC, Lev VL (eds) Membrane transport in red cells. Academic, New York, pp 137–172

Legrum B, Fasold H, Passow H (1980) Enhancement of anion equilibrium exchange by dansylation of the red blood cell membrane. Hoppe-Seyler's Z Physiol Chem 361: 1573–1590

Lepke S, Passow H (1976) Effects of incorporated trypsin on anion exchange and membrane proteins in human red blood cell ghosts. Biochim Biophys Acta 455: 353–370

Lepke S, Passow H (1982) Inverse effects of dansylation of the red blood cell membrane on band 3 protein-mediated transport of sulphate and chloride. J Physiol 328:27–48

Lepke S, Fasold H, Pring M, Passow H (1976) A study of relationship between inhibition of anion exchange and binding to the red blood cell membrane of 4,4′-diisothiocyano-stilbene-2,2′-disulfonic acid (DIDS) and of its dihydro derivative (H_2DIDS). J Membr Biol 29:147–177

Levinson C (1982) Chloride transport in the Ehrlich mouse ascites tumor cell. In: Zadunaisky JA (ed) Chloride transport in biological membranes. Academic, New York, pp 383–396

Lieb WR, Stein WD (1972) Carrier and non-carrier models for sugar transport in the human red blood cell. Biochim Biophys Acta 265:187–207

Lieberman DM, Reithmeier RAF (1983) Characterization of the stilbene disulfonate binding site of band 3 polypeptide of human erythrocyte membranes. Biochemistry 22:4028–4033

Lodish H, Braell WA (1982) Synthesis and maturation of the erythrocyte anion transport protein and internal sequence for membrane insertion. Biochem Soc Symp 47: 193–209

Löw I, Friedrich T, Burckhardt G (1984) Properties of an anion exchanger in rat renal basolateral membrane vesicles. Am J Physiol 246:F334–F342

Low PS (1978) Specific cation modulation of anion transport across the human erythrocyte membrane. Biochim Biophys Acta 514:264–273

Low PS, Westfall MA, Allen DP, Appell KC (1984) Characterization of the reversible conformational equilibrium of the cytoplasmic domain of erythrocyte membrane band 3. J Biol Chem 259:13070–13076

Low PS, Waugh SM, Zinke K, Drenckhahn D (1985) The role of hemoglobin denaturaion and band 3 clustering in red blood cell aging. Science 227:531–533

Lu YB, Chow EH (1982) Bicarbonate/chloride transport kinetics at 37°C and its relationship to membrane lipids in mammalian erythrocytes. Biochim Biophys Acta 689:485–489

Lukacovic MF, Feinstein MB, Sha'afi RI, Perrie S (1981) Purification of stabilized band 3 protein of the human erythrocyte membrane and its reconstitution into liposomes. Biochemistry 20:3145–3151

Lukacovic MF, Toon MR, Solomon AK (1984a) Site of red cell cation leak induced by mercurial sulphydryl reagents. Biochim Biophys Acta 772:313–320

Lukacovic MF, Verkman AS, Dix JA, Solomon AK (1984b) Specific interaction of the water transport inhibitor, pCMBS, with band 3 in red blood cell membranes. Biochim Biophys Acta 778:253–259

Lysko KA, Carlson R, Taverna R, Snow J, Brandts JF (1981) Protein involvement in structural transitions of erythrocyte ghosts. Use of thermal gel analysis to detect protein aggregation. Biochemistry 20:5570–5576

Macara IG, Cantley LC (1981a) Interactions between transport inhibitors at the anion binding sites of the band 3 dimer. Biochemistry 20:5095–5105

Macara IG, Cantley LC (1981b) Mechanism of anion exchange across the red cell membrane by band 3: interactions between stilbene-disulfonate and NAP-taurine binding sites. Biochemistry 20:5695–5701

Macara IG, Cantley LC (1983) The structure and function of band 3. In: Elson E, Frazier W, Glaser L (eds) Cell membranes: methods and reviews, vol I. Plenum, NEw York, pp 47–87

Macara IG, Kuo S, Cantley LC (1983) Evidence that inhibitors of anion exchange induce a transmembrane conformational change in band 3. J Biol Chem 258:1785–1792

Maddy AH (1964) A fluorescent label for the outer components of the plasma membrane. Biochim Biophys Acta 88:390–399

Maretzki D, Groth J, Tsamaloukas AG, Gründel M, Krüger S, Rapoport S (1974) The membrane association and dissociation of human glyceraldehyde-3-phosphate dehydrogenase under various conditions of hemolysis. Immunochemical evidence for the lack of binding under cellular conditions. FEBS Lett 39:83–87

Markowitz S, Marchesi VT (1981) The cyrboxyl-terminal domain of human erythrocyte band 3. Description, isolation and location in the bilayer. J Biol Chem 256: 6463–6468

Matsuyama H, Kawano Y, Hamasaki N (1983) Anion transport activity in the human erythrocyte membrane modulated by proteolytic digestion of the 38,000-dalton fragment in band 3. J Biol Chem 258:15376–15381

Mawby WJ, Findlay JB (1982) Characterization and partial sequence of di-iodosulphophenyl isothiocyanate-binding peptide from human erythrocyte anion-transport protein. Biochem J 205:465–475

Mercer RW, Dunham PB (1982) Membrane-bound ATP fuels the Na/K pump. Studies on membrane-bound glycolytic enzymes on inside-out vesicles from human red cell membranes. J Gen Physiol 78:547–568

Mikkelsen RB, Wallach DFH (1976) Photoactivated cross-linking of proteins within the erythrocyte membrane core. J Biol Chem 251:7413–7416

Milanick MA (1980) Ordered reaction of sulfate and protons with the anion exchange mechanism of human red blood cells. Fed Proc 39:1715

Milanick MA, Gunn RB (1982a) Proton-sulfate co-transport: mechanism of H^+ and sulfate addition to the chloride transporter of human red blood cells. J Gen Physiol 79:87–113

Milanick MA, Gunn RB (1982b) Interactions between external protons and the anion transporter of human erythrocytes. Biophys J 37:213a

Milanick MA, Gunn RB (1984) Proton-sulfate cotransport: external proton activation of sulfate influx into human red blood cells. Am J Physiol 247:C247–C259

Miller C (1984) Ion channels in liposomes. Ann Rev Physiol 46:549–558

Mond R (1927) Umkehr der Anionenpermeabilität der roten Blutkörperchen in eine selektive Permeabilität der Kationen. Pflüger's Arch Ges Physiol 217:618–630

Morgan M, Hanke P, Grygorczik R, Tintschl A, Fasold H, Passow H (1985) Mediation of anion transport in oocytes of Xenopus laevis by biosynthetically inserted band 3 protein from mouse spleen erythroid cells. EMBO J 4:1927–1931

Motais R, Cousin JL (1978) A structure activity study of some drugs acting as reversible inhibitors of chloride permeability in red cell membranes: influence of ring substituents. In: Straub RW, Bolis L (eds) Cell membrane receptors for drugs and hormones: A multidisciplinary approach. Raven, New York, pp 219:225

Mühlebach T, Cherry RJ (1982) Influence of cholesterol on the rotation and self-association of band 3 in the human erythrocyte membrane. Biochem 21:4225–4228

Mueller TJ, Morrison M (1977) Detection of a variant of protein 3, the major transmembrane protein of the human erythrocyte. J Biol Chem 252:6573–6576

Mueller TJ, Li YT, Morrison M (1979) Effect of endo-β-galactosidase on intact huamn erythrocytes. J Biol Chem 254:8103–8106

Muirhead KA, Steinfeldt RC, Severski MC, Knauf PA (1984) Anion transport heterogeneity detected by flow cytometric measurement of NBD-taurine efflux kinetics. Cytometry 5:268–274

Murthy SNP, Liu T, Köhler H, Steck TL (1981a) The aldolase binding site of the human erythrocyte membrane. Primary structure of the amino-terminal decapeptide of band 3. J Supramol Struct (Suppl) 5:125

Murthy SNP, Liu T, Kaul RK, Köhler H, Steck TL (1981b) The aldolase-binding site of the human erythrocyte membrane is at the NH_2 terminus of band 3. J Biol Chem 256:11203–11208

Murthy SNP, Kaul RK, Köhler H (1984) Hemoglobin binds to the amino-terminal 23-residue fragment of human erythrocyte band 3 protein. Hoppe-seyler's Z Physiol Chem 365:9–17

Nanri H, Hamasaki N, Minakami S (1983) Affinity labeling of erythrocyte band 3 proteins with pyridoxal 5-phosphate involvement of the 35,000 dalton fragment in anion transport. J Biol Chem 258:5985–5989

Nigg EA, Cherry RJ (1979) Influence of temperature and cholesterol on the rotational diffusion of band 3 in the human erythrocyte membrane. Biochemistry 18:3457–3465

Nigg EA, Cherry RJ (1980) Anchorage of a band 3 population at the erythrocyte cytoplasmic membrane surface: protein rotational diffusion measurements. Proc Natl Acad Sci 77:4702–4706

Obaid AL, McElroy-Critz A, Crandall ED (1979) Kinetics of bicarbonate/chloride exchange in dogfish erythrocytes. Am J Physiol 237:132–138

O'Connor CM, Clarke S (1983) Methylation of erythrocyte membrane proteins at extracellular and intracellular D-aspartyl sites in vitro. J Biol Chem 258:8485–8492

Pappert G, Schubert D (1982) Self-association of band 3 protein from erythrocyte membranes in solutions of a non-ionic detergent, Ammonyx-LO. Protides Biol Fluids 29:117–121

Pappert G, Schubert D (1983) The state of association of band 3 protein of the human erythrocyte membrane in solutions of nonionic detergents. Biochim Biophys Acta 730:32–40

Passow H (1969) Passive ion permeability of the erythrocyte membrane. Prog Biophys Mol Biol 19:425–467

Passow H (1971) Effects of pronase on passive ion permeability of the human red blood cell. J Membr Biol 6:233–258

Passow H (1978) The binding of 1-fluoro-2,4-dinitrobenzene and of certain stilbene-2,2′-disulfonic acids to anion permeability-controlling sites on the protein in band 3 of the red blood cell membrane. In: Straub RW, Bolis L (eds) Cell membrane receptors for drugs and hormones: a multidisciplinary approach. Raven, New York, pp 203–218

Passow H (1982) Anion-transport-related conformational changes of the band 3 protein in the red blood cell membranes. In: Martonosi AN (ed) Membranes and transport, vol II. Plenum, New York, pp 451–460

Passow H, Fasold H (1980) On the mechanism of band-3-protein mediated anion transport across the red blood cell membrane. In: Hollan G, Gardos G, Sarkadi B (eds) Genetics, structure and function of blood cells. Pergamon/Akademiai Kiado, Budapest, pp 249–261 (Proc 28th Intern Congr of Physiol Sciences, Budapest)

Passow H, Schnell KF (1969) Chemical modifiers of passive ion permeability of the erythrocyte membrane. Experientia 25:460–468

Passow H, Wood PG (1974) Current concepts of the mechanism of anion permeability. In: Callingham BA (ed) Drugs and transport processes. McMillan, London, pp 149–171

Passow H, Zaki L (1978) Studies on the molecular mechanism of anion transport across the red blood cell membrane. In: Karnovsky M (ed) Molecular specification and symmetry in membrane function. Harvard University Press, Cambridge, pp 229–252

Passow H, Fasold H, Zaki L, Schumann B, Lepke S (1975) Membrane proteins and anion exchange in human erythrocytes. In: Gardos G, Szasz I (eds) Biomembranes: structure and function. North Holland, Amsterdam, pp 197–214 (Proc 9th FEBS meeting Budapest)

Passow H, Fasold H, Lepke S, Pring M, Schumann B (1977) Chemical and enzymatic modification of membrane proteins and anion transport in human red blood cells. In: Miller MW, Shamoo AE (eds) Membrane toxicity. Plenum, New York, pp 353–379

Passow H, Fasold H, Gärtner M, Legrum B, Ruffing W, Zaki L (1980a) Anion transport across the red blood cell membrane and the conformation of the protein in band 3. Ann NY Acad Sci 341:361–383

Passow H, Kampmann L, Fasold H, Jennings M, Lepke S (1980b) Mediation of anion transport across the red blood cell membrane by means of conformational changes of the band 3 protein. In: Lassen UV, Ussing HH, Wieth JO (eds) Membrane transport in erythrocytes. Munksgaard, Copenhagen, pp 345–367 (Alfred Benzon Symposium 14)

Passow H, Fasold H, Jennings ML, Lepke S (1982) The study of the anion transport protein ('band 3 protein') in the red cell membrane by means of tritiated 4,4′-diisothiocyano-dihydrostilbene-2,2′-disulfonic acid ($^3H_2$DIDS). In: Zadunaisky J (ed) Chloride transport in biological membranes. Academic, New York, pp 1–31

Passow H, Berghout A, Romano L (1984a) Band 3 protein mediated non electrogenic proton equilibration across the membranes of the red blood cells of mammals, amphibians and fish. In: Bolis L, Helmreich E, Passow H (eds) Information and energy transduction in biological membranes. Liss, New York, pp 95–102

Passow H, Ruffing W, Gärtner E, Legrum B (1984b) Modification of anion transport and the protein in band 3 by dinitrophenylation of the red blood cell membrane in chloride and sulfate media. Hoppe-Seyler's Z Physiol Chem 365:1041

Passow H, Shields M, LaCelle P, Grygorczyk R, Schwarz W, Peters R (1986) Effects of calcium on structure and function of the red blood cell membrane. In: Clarkson T (ed) The cytoskeleton: a target for toxic agents. Plenum. In press

Patlak CS (1957) Contributions to the theory of active transport: II The gate type non-carrier mechanism and generalisations concerning tracer flow, efficiency and measurement of the energy expenditure. Bull Math Biophys 19:209–235

Peters R (1981) Translational diffusion in the plasma membrane of single cells as studied by fluorescence microphotolysis. Cell Biol Int Rep 5:733–760

Peters R, Passow H (1984) Anion transport in single erythrocyte ghosts measured by fluorescence microphoto-lysis. Biochim Biophys Acta 777:334–338

Peters R, Peters J, Tews KH, Bahr W (1974) A microfluorometric study of translational diffusion in erythrocyte membranes. Biochim Biophys Acta 367:282–294

Pinto da Silva P (1972) Translational mobility of membrane intercalated particles of human erythrocyte ghosts. pH-dependent, reversible aggregation. J Cell Biol 53: 777–787

Raida M, Passow H (1985) Enhancement of divalent anion transport across the human red blood cell membrane by the water-soluble dansyl chloride derivative 2-(*N*-piperidine)ethylamine-1-naphtyl-5-sulfonylchloride (PENS-Cl). Biochim Biophys Acta 812:624–632

Ramjeesingh M, Gaarn A, Rothstein A (1980a) The location of a disulfonic stilbene binding site in band 3, the anion transport protein of the red blood cell membrane. Biochim Biophys Acta 599:127–139

Ramjeesingh M, Grinstein S, Rothstein A (1980b) Intrinsic segments of band 3 that are associated with anion transport across the red blood cell membranes. J Membr Biol 57:95–102

Ramjeesingh M, Gaarn A, Rothstein A (1981) The amino acid conjugate formed by the interaction of the anion transport inhibitor, DIDS, with band 3 protein from human red blood cell membranes. Biochim Biophys Acta 641:173–182

Ramjeesingh M, Gaarn A, Rothstein A (1982) The sulfhydryl groups of the 35,000-dalton C-terminal segment of band 3 are located in a 9000-dalton fragment produced by chymotrypsin treatment of red cell ghosts. J Bioenerg Biomembr 13:411–423

Ramjeesingh M, Gaarn A, Rothstein A (1983) The location of the three cysteine residues in the primary structure of the intrinsic segments of band 3 protein, and implications concerning the arrangement of band 3 protein in the bilayer. Biochim Biophys Acta 729:150–160

Ramjeesingh M, Gaarn A, Rothstein A (1984) Pepsin cleavage of band 3 produces its membrane-crossing domains. Biochim Biophys Acta 769:381–389

Rao A (1979) Disposition of the band 3 polypeptide in the human erythrocyte membrane. J Biol Chem 254:3503–3511

Rao A, Reithmeier RAF (1978) Reactive sulfhydryl groups of the band 3 polypeptide from human erythrocyte membrane. J Biol Chem 254:6144–6150

Rao A, Martin P, Reithmeier RAF, Cantley LC (1979) Location of the stilbene disulfonate binding site of the human erythrocyte anion-exchange system by resonance energy transfer. Biochemistry 18:4505–4516

Reithmeier RAF (1983) Inhibition of anion transport in human red blood cells by 5,5′-dithiobis(2-nitrobenzoic acid). Biochim Biophys Acta 732:122–125

Reithmeier RAF, Rao A (1979) Reactive sulfhydryl groups of the band 3 polypeptide from human erythrocyte membranes. Identification of the sulfhydryl groups involved in Cu^{2+}-*o*-phenanthroline cross-linking. J Biol Chem 254:6151–6155

Romano L, Passow H (1984) Characterization of the anion transport system in trout red blood cell. Am J Physiol 246:C330–C338

Ross AH, McConnell HM (1978) Reconstitution of the erythrocyte anion channel. J Biol Chem 253:4777–4782

Rothstein A (1982) Functional structure of band 3, the anion transport protein in the red blood cell, as determined by proteolytic and chemical cleavage. In: Martonosi AN (ed) Membranes and transport, vol II. Plenum, New York, pp 435–440

Rothstein A (1984) The functional architecture of band 3, the anion transport protein of the red cell membrane. Can J Biochem Cell Biol 62:1198–1204

Rothstein A, Cabantchik ZI, Knauf P (1976) Mechanism of anion transport in red blood cells: role of membrane proteins. Fed Proc 35:3–10

Rudloff V, Lepke S, Passow H (1983) Inhibition of anion transport across the red cell membrane by dinitrophenylation of a specific lysine residue at the H_2DIDS binding site of the band 3 protein. FEBS Letters 163:14–21

Runyon KR, Gunn RB (1984) Phosphate-chloride exchange in human red blood cells: monovalent vs divalent phosphate transport. Biophys J 45:18a

Sabatini DD, Kreibich G, Morimoto T, Adesnik M (1982) Mechanisms for the incorporation of proteins in membranes and organelles. J Cell Biol 92:1–22

Sabban EL, Sabatini DD, Marchesi VT, Adesnik M (1980) Biosynthesis of erythrocyte membrane protein band 3 in DMSO-induced friend erythroleukemia cells. J Cell Physiol 104:261–268

Sabban E, Marchesi V, Adesnik M, Sabatini DD (1981) Erythrocyte membrane protein band 3: its biosynthesis and incorporation into membranes. J Cell Biol 91:637–646

Saffmann PG, Delbrück M (1975) Brownian motion in biological membranes. Proc Natl Acad Sci 72:3111–3113

Saleemuddin M, Zimmermann U, Schneeweiß F (1977) Preparation of human erythrocyte ghosts in isotonic solution: haemoglobin content and polypeptide composition. Z. Naturforsch 32C:627–631
Salhany JM, Rauenbuehler PR (1983) Kinetics and mechanism of erythrocyte anion exchange. J Biol Chem 258:245–249
Salhany JM, Shaklai N (1979) Functional properties of human hemoglobin bound to the erythrocyte membrane. Biochemistry 18:893–899
Salhany JM, Cordes KA, Gaines ED (1980) Light-scattering measurements of hemoglobin binding to the erythrocyte membrane. Evidence for transmembrane effects related to a disulfonic stilbene binding to band 3. Biochemistry 19:1447–1454
Scarpa A, Cecchetto A, Azzone GF (1970) The mechanism of anion translocation and pH equilibration in erythrocytes. Biochim Biophys Acta 219:179–188
Scheuring U, Kollewe K, Schubert D (1984) A new method for the reconstitution of the anion transport system of the human erythrocyte membrane. Hoppe-Seyler's Z Physiol Chem 365:1056–1057
Scheuring U, Kollewe K, Haase W, Schubert D (to be published) A new method for the reconstitution of the anion transport system of the human erythrocyte membrane. J Membrane Biol
Schnell KF (1977) Anion transport across the red blood cell membrane mediated by dielectric pores. J Membr Biol 37:99–136
Schnell KF, Besl E (1984) Concentration dependence of the unidirectional sulfate and phosphate flux in human red cell ghosts under self-exchange and under homoexchange conditions. Pflügers Arch 402:197–206
Schnell KF, Gerhardt S, Schöppe-Fredenburg A (1977) Kinetic characteristics of the sulfate self-exchange in human red blood cells and red blood cell ghosts. J Membr Biol 30:319–350
Schnell KF, Besl E, v der Mosel R (1981) Phosphate transport in human red blood cells: concentration dependence and pH dependence of the unidirectional phosphate flux at equilibrium conditions. J Membr Biol 61:173–192
Schnell KF, Elbe W, Käsbauer J, Kaufmann E (1983) Electron spin resonance studies on the inorganic anion transport system of the human red blood cell binding of a disulfonate stilbene spin label (NDS-Tempo) and inhibition of anion transport. Biochim Biophys Acta 732:266–275
Schubert D, Boss K (1982) Band 3 protein-cholesterol interactions in erythrocyte membranes. Possible role in anion transport and dependency on membrane phospholipid. FEBS Lett 150:4–8
Schubert D, Domning B (1978) A new method for the preparation of band 3, the main integral protein of the human erythrocyte membrane. Hoppe-Seyler's Z Physiol Chem 359:507–515
Schubert D, Boss K, Dorst HJ, Flossdorf J, Pappert G (1983) The nature of the stable noncovalent dimers of band 3 protein from erythrocyte membranes in solutions of Triton X-100. FEBS Lett 163:81–84
Schwoch G, Rudloff V, Wood-Guth I, Passow H (1974) Effect of temperature on sulfate movements across chemically or enzymatically modified membranes of human red blood cells. Biochim Biophys Acta 339:126–138
Sha'afi RI, Feinstein MB (1977) Membrane water channels and SH groups. In: Miller MW, Shamoo AE (eds) Membrane toxicity. Plenum, New York, pp 67–83
Shami Y, Carver J, Ship S, Rothstein A (1977) Inhibition of Cl^- binding to anion transport protein of the red blood cell by DIDS (4,4'-diisothiocyno-2,2'-stilbenedisulfonic acid) measured by (^{35}Cl) NMR. Biochim Biophys Res Comm 76:429–436
Shami Y, Rothstein A, Knauf PA (1978) Identification of the Cl^- transport site of human blood cells by a kinetic analysis of the inhibitory effects of a chemical probe. Biochim Biophys Acta 508:357–363
Shanahan MF, D'Artel-Ellis J (1984) Orientation of the glucose transporter in the human erythrocyte membrane. Investigation by in situ proteolytic dissection. J Biol Chem 259:13878–13884

Shanahan MF (1982) A natural photoaffinity ligand for labeling the human erythrocyte glucose transporter. J Biol Chem 257:7290–7293

Shaklai N, Yguerabide J, Ranney HM (1977) Classification and localization of hemoglobin binding sites on the red blood cell membrane. Biochemistry 16:5593–5597

Sheetz M, Schindler M, Koppel DE (1980) Lateral mobility of integral membrane proteins is increased in spherocytic erythrocytes. Nature 285:510–512

Shelton RL, Langdon RG (1983) Reconstitution of glucose transport using human erythrocyte band 3. Biochim Biophys Acta 733:25–33

Shinitzky M, Rivnay B (1977) Degree of exposure of membrane proteins determined by fluorescence quenching. Biochemistry 16:982–986

Ship S, Shami Y, Breuer W, Rothstein A (1977) Synthesis of tritiated (^{3}H)DIDS and its covalent reaction with sites related to anion transport in red blood cells. J Membr Biol 33:311–324

Sigrist H, Zahler P (1982) Hydrophobic labeling and cross-linking of membrane proteins. In: Martonosi AN (ed) Membranes and transport, vol I. Plenum, New York, pp 173–184

Sigrist H, Kempf C, Zahler P (1980) Interaction of phenylisothiocyanate with human erythrocyte band 3. I. Covalent modification and inhibition of phosphate transport. Biochim Biophys Acta 597:137–144

Simmons NL (1981) Stimulation of Cl^- secretion by exogenous ATP in cultured MDCK epithelial monolayers. Biochim Biophys Acta 646:231–242

Singer J, Morrison M (1980) Effect of adenosine on concanavalin A agglutination of human erythrocytes. Biochim Biophys Acta 598:40–50

Smith PL, Orellana SA, Field M (1981) Active sulfate absorption in rabbit ileum: dependence on sodium and chloride and effects of agents that alter chloride transport. J Membr Biol 63:199–206

Snow JW, Vincentelli J, Brandts JF (1981) A relationship between anion transport and a structural transition of the human erythrocyte membrane. Biochim Biophys Acta 642:418–428

Solomon AK, Chasan B, Dix JA, Lukacovic MF, Toon MR, Verkman AS (1982) The aqueous pore in the red cell membrane: band 3 as a channel for anions, cations, non electrolytes and water. Biophys J 37:215a

Solomon AK, Chasan B, Dix JA, Lukacovic MF, Toon MR, Verkman AS (1983) The aqueous pore in the red cell membrane: band 3 as a channel for anions, cations, non-electrolytes, and water. Ann NY Acad Sci 414:97–124

Staros JV, Kakkad BP (1983) Cross-linking and chymotryptic digestion of the extra cytoplasmic domain of the anion exchange channel in intact human erythrocytes. J Membr Biol 74:247–254

Steck TL (1972) Cross-linking the mjaor proteins of the isolated erythrocyte membrane. J Mol Biol 66:295–305

Steck TL (1974) The organization of proteins in the human red cell membrane. J Cell Biol 62:1–19

Steck TL (1978) The band 3 protein of the human red cell membrane: a review. J Supramol Struct 8:311–324

Steck TL, Yu J (1973) The selective solubilization of proteins from red blood cell membranes by protein pertubants. J Supramol Struct 1:220–232

Steck TL, Ramos B, Strapazon E (1976) Proteolytic dissection of band 3, the predominant transmembrane polypeptide of the human erythrocyte membrane. Biochemistry 15:1154–1161

Steck TL, Koziarz JJ, Singh MK, Reddy G, Köhler H (1978) Preparation and analysis of seven major, topographically defined fragments of band 3, the predominant transmembrane polypeptide of human erythrocyte membranes. Biochemistry 17: 1216–1222

Strapazon E, Steck TL (1976) Binding of rabbit muscle aldolase to band 3, the predominant polypeptide of the human erythrocyte membrane. Biochemistry 15: 1421–1424

Strapazon E, Steck TL (1977) Interaction of the aldolase and the membrane of human erythrocytes. Biochemistry 16:2966–2971

Tanford C (1962) The interpretation of hydrogen ion titration curves of proteins. Adv Protein Chem 17:69–165

Tanford C (1985) Simple model can explain self-inhibition of red cell anion exchange. Biophys J 47:15–20

Tanner MJA (1979) Isolation of integral membrane proteins and criteria for identifying carrier proteins. Curr Top Membr Transp 12:1–51

Tanner MJA, Anstee DJ (1976) A method for the direct demonstration of the lectin-binding components of the human erythrocyte membrane. Biochem J 153:265–270

Tanner MJA, Williams DG, Jenkins RE (1980) Structure of the erythrocyte anion transport proteins. Ann NY Acad Sci 341:455–464

Tosteson DC, Gunn RB, Wieth JO (1973) Chloride and hydroxyl ion conductance of sheep red cell membrane. In: Gerlach E, Moser K, Deutsch E, Wilmanns W (eds) Erythrocytes, thrombocytes, leucocytes. Thieme, Stuttgart, pp 62–69

Tsai IH, Murthy SP, Steck TL (1982) Effect of red cell membrane binding on the catalytic activity of glyceraldehyde-3-phosphate dehydrogenase. J Biol Chem 257: 1438–1442

Tsuji T, Irimura T, Osawa T (1980) The carbohydrate moiety of band 3 glycoprotein of human erythrocyte membranes. Biochem J 187:677–686

Tsuji T, Irimura T, Osawa T (1981) The carbohydrate moiety of band 3 glycoprotein of human erythrocyte membranes. Structure of lower molecular weight oligosaccharides. J Biol Chem 256:10497–10502

Van Hoogevest P, Van Duijn G, Batenburg AM, De Kruijff B, De Gier J (1983) The anion permeability of vesicles reconstituted with intrinsic proteins from the human erythrocyte membrane. Biochim Biophys Acta 734:1–17

Verkman AS, Dix JA, Solomon AK (1981) Thermodynamics of stilbene binding sites on human red cell band 3. Biophys J 33:48a

Verkman AS, Dix JA, Solomon AK (1982) A non-competitive 'shunt' pathway for the effect of chloride on the band 3-DBDS conformational change in red cell membranes. Biophys J 37:216a

Verkman AS, Dix JA, Solomon AK (1983) Anion transport inhibitor binding to band 3 in red blood cell membranes. J Gen Physiol 81:421–449

Wang K, Richards FM (1974) An approach to nearest neighbour analysis of membrane proteins. J Biol Chem 249:8005–8018

Wang K, Richards FM (1975) Reaction of dimethyl-3,3′-dithiobispropionimidate with intact human erythrocytes. Cross-linking of membrane proteins and hemoglobin. J Biol Chem 250:6622–6626

Weinstein RS, Khodadad JK, Steck TL (1978) Fine structure of the band 3 protein in human red cell membranes: freeze-fracture studies. J Supramol Struct 8:325–335

Weinstein RS, Khodadad JK, Steck TL (1980) The band 3 protein intramembrane particle of the human red blood cell. In: Lassen UV, Ussing HH, Wieth JO (eds) Membrane transport in erythrocytes. Munksgaard, Copenhagen, pp 35–48 (Alfred Benzon Symposium 14)

Wiedemann B, Elbaum D (1983) Effect of hemoglobin A and S in human erythrocyte ghosts. J Biol Chem 258:5483–5489

Wieth JO (1979) Bicarbonate exchange through the human red cell membrane determined with ^{14}C-bicarbonate. J Physiol 294:521–539

Wieth JO, Bjerrum PJ (1982) Titration of transport and modifier sites in the red cell anion transport system. J Gen Physiol 79:253–282

Wieth JO, Bjerrum PJ (1983) Transport and modifier sites in capnophorin, the anion transport protein of the erythrocyte membrane. In: Quagliariello E, Palmieri F (eds) Structure and function of membrane proteins. Elsevier, Amsterdam, pp 95–106

Wieth JO, Brahm J (1980) Kinetics of bicarbonate exchange in human red cells – physiological implications. In: Lassen UV, Ussing HH, Wieth JO (eds) Membrane transport in erythrocytes. Munksgaard, Copenhagen, pp 467–482 (Alfred Benzon Symposium 14)

Wieth JO, Brahm J (1985) Cellular anion transport. In: Seldin DW, Giebisch G (eds) Physiology and pathophysiology. The kidney, vol I. Raven, New York, pp 49–89

Wieth JO, Brahm J, Funder J (1980) Transport and interactions of anions and protons in the red blood cell membrane. Ann NY Acad Sci 341:394–418

Wieth JO, Andersen OS, Brahm J, Bjerrum PJ, Borders CL (1982a) Chloride-bicarbonate exchange in red blood cells. Philos Trans R Soc Lond (Biol) 299:383–399

Wieth JO, Bjerrum PJ, Andersen OS (1982b) The anion transport protein of the red cell membrane. A zipper mechanism of anion exchange. Tokai J Exp Clin Med 7: 91–101

Wieth JO, Bjerrum PJ, Borders CL (1982c) Irreversible inactivation of red cell chloride exchange with phenylgloxal, an arginine-specific reagent. J Gen Physiol 79:283–312

Williams DG, Jenkins RE, Tanner MJA (1979) Structure of the anion-transport protein of the human erythrocyte membrane. Further studies on the fragment produced by proteolytic digestion. Biochem J 181:477–493

Yeltman DR, Harris BG (1980) Localization and membrane association of aldolase in human erythrocytes. Arch Biochem Biophys 199:186–196

Yoon SC, Toon MR, Solomon AK (1984) Relation between red cell anion exchange and water transport. Biochim Biophys Acta 778:385–389

Yu J, Steck TL (1975) Associations of band 3, the predominant polypeptide of the human erythrocyte membrane. J Biol Chem 250:9170–9175

Yu J, Fischman DA, Steck TL (1973) Selective solubilization of proteins and phospholipids from red cell membranes by non-ionic detergents. J Supramol Struct 1:233–248

Zaki L (1981) Inhibition of anion transport across red blood cells with 1,2-cyclohexanedione. Biochem Biophys Res Comm 99:243–251

Zaki L (1982) The effect of arginine specific reagents on anion transport across the red blood cells. Protides Biol Fluids 29:279–282

Zaki L (1983) Anion transport in red blood cells and arginine specific reagents. (1) Effect of chloride and sulfate ions on phenylglyoxal sensitive sites in the red blood cell membrane. Biochem Biophys Res Comm 110:616–624

Zaki L (1984) Anion transport in red blood cells and arginine-specific reagents. The location of ^{14}C phenylglyoxal binding sites in the anion transport protein in the membrane of human red cells. FEBS Lett 169:234–240

Zaki L, Julien T (1983) Inactivation of the anion transport system in the red blood cell membrane by α-dicarbonyl reagents. Hoppe-Seyler's Z Physiol Chem 364:1233

Zaki L, Julien T (1985) Anion transport in red blood cells and arginine specific reagents. Interaction between the substrate binding site and the binding site of arginine specific reagents. Biochim Biophys Acta 818:325–332

Zaki L, Fasold H, Schumann B, Passow H (1975) Chemical modification of membrane proteins in relation to inhibition of anion exchange in human red blood cells. J Cell Physiol 36:471–494

Zwaal RF, Roelofsen B, Colley CM (1973) Localisation of red cell membrane constituents. Biochim Biophys Acta 300:159–182

Author Index

Subject Index

GPSR Compliance

The European Union's (EU) General Product Safety Regulation (GPSR) is a set of rules that requires consumer products to be safe and our obligations to ensure this.

If you have any concerns about our products, you can contact us on ProductSafety@springernature.com

In case Publisher is established outside the EU, the EU authorized representative is:

Springer Nature Customer Service Center GmbH
Europaplatz 3
69115 Heidelberg, Germany

Batch number: 10620881

Printed by Printforce, the Netherlands